BCg

Ground-based optical astronomy is undergoing a renaissance inspired by recent technological advances. This book collects contributions from expert international designers and observers, which show the fundamental role of innovative optical design in modern astrophysics and cosmological research. The papers are based on those presented at a conference held at the Royal Greenwich Observatory, Cambridge, in September 1991 in celebration of Professor C. G. Wynne. New and future telescopes, instruments, spectrographs and detector-systems are described in the context of the research problems at which they are targeted. This book is an essential reference for both state-of-the-art instrumentation and future design.

Optics in Astronomy

32nd Herstmonceux Conference 9–11 September 1991
back row: I. Escudero-Sanz, N.M. Parker, G. Gilmore, M. Bridgeland, J.D.H. Pilkington,
J. Maxwell, I.F. Corbett, J.R. Walsh, M. Rosete-Agular, R.E.S. Clegg
middle row: Y.S. Kim, J.C. Dainty, A.K. Forrest, M. Wells, J.A. Bailey, P.R. Gillingham,
M. Pettini, K. Taylor, G.M. Sisson, R.V. Willstrop, E.J. Hysom, G.E. Satterthwaite, J. Meaburn
front row: R.C.M. Learner, R.G. Bingham, M.C. Morris, T.J. Lee, C.R. Kitchin, C. Roddier,
G.C. Wynne, F. Graham-Smith, A. Boksenberg, J.V. Wall, R. Racine,
R. Hanbury Brown, P.A. Wayman and C.D. McKeith.

Optics in Astronomy

Edited by
J. V. Wall
Royal Greenwich Observatory, Cambridge

CAMBRIDGE
UNIVERSITY PRESS

Published by the Press Syndicate of the University of Cambridge
The Pitt Building, Trumpington Street, Cambridge CB2 1RP
40 West 20th Street, New York, NY 10011–4211, USA
10 Stamford Road, Oakleigh, Melbourne 3166, Australia

© Cambridge University Press 1993

First published 1993

Printed in Great Britain at the University Press, Cambridge

A catalogue record for this book is available from the British Library

Library of Congress cataloguing in publication data available

ISBN 0 521 44511 6 hardback

Contents

CONTENTS

Participants

R. W. ARGYLE, *Royal Greenwich Observatory*
J. A. BAILEY, *Anglo-Australian Observatory*
A. BOKSENBERG, *Royal Greenwich Observatory*
R. G. BINGHAM, *Royal Greenwich Observatory*
R. HANBURY BROWN
M. T. BRIDGELAND, *Royal Greenwich Observatory*
R. E. S. CLEGG, *Royal Greenwich Observatory*
I. F. CORBETT, *Science and Engineering Research Council*
J. C. DAINTY, *Imperial College London*
D. W. DEWHIRST, *Institute of Astronomy*
I. ESCUDERO-SANZ, *Institute of Astronomy*
A. K. FORREST, *Imperial College London*
D. W. GELLATLY, *Royal Greenwich Observatory*
P. R. GILLINGHAM, *Anglo-Australian Observatory*
G. F. GILMORE, *Institute of Astronomy*
F. GRAHAM-SMITH, *Nuffield Radio Astronomy Laboratories*
W. HAN, *University College London*
E. J. HYSOM, *A E Optics, Cambridge*
C. R. JENKINS, *Royal Greenwich Observatory*
P. R. JORDEN, *Royal Greenwich Observatory*
Y-S. KIM, *University College London*
C. R. KITCHIN, *Hatfield Polytechnic Observatory*
R. C. M. LEARNER, *Imperial College London*
T. J. LEE, *Royal Observatory Edinburgh*
J. V. MAJOR, *University of Durham*
J. MAXWELL, *Imperial College London*
C. D. McKEITH, *Queens University Belfast*
D. McMULLAN, *Cavendish Laboratory*
J. MEABURN, *University of Manchester*
F. MERKLE, *European Southern Observatory Garching*
M. C. MORRIS, *Rutherford Appleton Laboratory*
N. M. PARKER, *Royal Greenwich Observatory*
M. PETTINI, *Royal Greenwich Observatory*
K. A. R. B. PIETRASZEWSKI, *Queensgate Instruments*
J. D. H. PILKINGTON, *Royal Greenwich Observatory*
R. RACINE, *Université de Montréal*
C. RODDIER, *University of Hawaii/CFHT*
M. ROSETE-AGUILAR, *Imperial College London*
G. E. SATTERTHWAITE, *Imperial College London*

LIST OF PARTICIPANTS

K. Sug-Whan, *University College London*
T. Shanks, *University of Durham*
G. M. Sisson
K. Taylor, *Anglo-Australian Observatory*
S. W. Unger, *Royal Greenwich Observatory*
D. D. Walker, *University College London*
J. V. Wall, *Royal Greenwich Observatory*
J. R. Walsh, *ST-ECF Garching*
P. J. Warner, *Mullard Radio Astronomy Observatory*
(F. G. Watson, *Royal Observatory Edinburgh*)
P. A. Wayman, *Dunsink Observatory* M. Wells, *Royal Observatory Edinburgh*
R. V. Willstrop, *Institute of Astronomy*
S. P. Worswick, *Royal Greenwich Observatory*
C. G. Wynne, *Institute of Astronomy*

Foreword

The Thirty-Second Herstmonceux Conference was held at the Royal Greenwich Observatory, Cambridge, 9–11 September 1991, in celebration of the 80th birthday of Professor C. G. Wynne, FRS.

Optics in Astronomy refers to a subject area whose range clearly exceeds the scope of any single scientific meeting, and exceeds the grasp of most of us associated with astronomy. There is a notable exception. Charles Wynne, to whom this conference and these proceedings are dedicated, has contributed immeasurably to our science, and he continues to do so. The extent of his contribution has been eloquently presented by Graham Smith and Gerry Gilmore whose contributions open and close this volume. Perhaps the contributions in between provide an even stronger testament to the influence of Charles Wynne – a tremendous range of instruments and observations, which almost invariably use photons traversing components which Charles Wynne has considered, conceived or designed himself.

I thank Martin Rees and Paul Murdin for their support in the initial consideration of this conference, Jasper Wall (the "Scientific Organizing Committee") for organizing it, and Monica Danforth (the "Local Organizing Committee") for doing so much of the work necessary to make it happen. I am very grateful to the participants for coming to honour our eminent colleague and for presenting such fine contributions. Finally, thank you Charles Wynne – we wouldn't have had this meeting without you, nor perhaps would we have had many of our prized astronomical observations. Many Happy Returns!

Alec Boksenberg
RGO, September 1991

I Introduction

1

An Optical Designer for UK Astronomy

F. Graham-Smith *

My main task when I was appointed to the RGO in 1974 was to organise the design and construction of the proposed Northern Hemisphere Observatory. At that time even the site was not determined, although the ROE team was already making encouraging remarks about the Canary Islands and particularly about La Palma. Alan Hunter, who took over as Director after Margaret Burbidge, had already made my task easier by providing key people such as John Pope, telescope designer, and George Harding, Project Scientist (fresh from moving the 74-in Radcliffe reflector[1] from Pretoria to Sutherland). There was also within the RGO considerable experience in the design and construction of spectrographs and photometers, which were used with the Isaac Newton Telescope and other telescopes at Herstmonceux and at Sutherland. Some of this experience served us well by indicating that up-to-date instrumentation is both very important and difficult to achieve. Hunter set out this situation for me when I arrived as Director Designate, and added almost as an afterthought that some help might be available from an optical designer who had taken early retirement and who had agreed to work for RGO on a part-time basis for a year or so. He was, of course, Charles Wynne. Seventeen years later Charles is still working for UK astronomy, and we are celebrating an undiminished stream of contributions to the design of telescopes and instruments, especially spectrographs.

Lecturing in optics in a university physics course did not qualify me to join the optics design team at RGO, but I had some general principles to help me understand their work. For example, the resolving power of a spectrograph can be understood simply through wavefront concepts which were already familiar in radio astronomy. In the very basic instrument of Fig. 1 a wave front is divided into two and recombined after reflection, with a difference D in path length. If in the combined beam the two parts are in phase, what change $\delta\lambda$ in wavelength is needed to put the two into antiphase, giving a minimum instead of a maximum intensity? We immediately find that the resolution $\delta\lambda/\lambda$ of this spectrograph is D/λ, the number of wavelengths in the extra path.

In Fig. 2 we advance to a grating, blazed and with a line spacing such that light returns along nearly the same path. The resolution is determined again by the path

*Nuffield Radio Astronomy Laboratories, Jodrell Bank, Macclesfield, Cheshire SK11 9DN, UK.
[1]See G.M. Sisson, this volume – *Ed.*

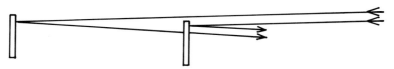

Figure 1. A two-element spectrograph, path-length difference *D*.

Figure 2. A grating blazed and with line spacing such that light returns along nearly the same path.

difference between the extreme edges of the wavefront, and in the limit by the width of the grating. If, as in Fig. 3, the grating is immersed in glass with a refractive index μ, and the line spacing is suitably adjusted, the resolving power is increased by a factor μ. I note with some pleasure that the immersed grating is the subject of some recent work by Charles, who incidentally shows how dangerous it is to oversimplify in this way.

So far it was easy for me to understand. The real problem comes in getting the light into the spectrograph, through the disperser, and on to the detector, without loss and covering a suitably wide angular range on the sky and a suitably wide range of wavelengths. My physics approach, using wavefront concepts, gave some limited help. I have a simple picture of how compressing a beam laterally is related to angular spread of rays; here an extra path length *d* corresponds to an angle inversely proportional to the width of the beam, allowing where appropriate for a ratio of refractive index. But in the end there is no help for it, the experts have to take over and do the geometric optics properly. Here the key advance, electronic digital computation applied to lens design, was described by Charles in a series of papers starting 1959.

Charles Wynne first became known among astronomers for his design of field correctors for the prime focus of telescopes. This work traces back to a 1949 paper on correctors for parabolic mirrors, and there has been a continuous stream

Figure 3. A grating immersed in glass of refractive index μ.

of designs of wide-field telescopes and field correctors. Perhaps the best-known to UK astronomers are the correctors for the Isaac Newton Telescope and the Anglo-Australian Telescope. An outstanding advance was the demonstration that spaced three-element correctors could perform as well as the earlier four-element correctors, both for paraboloids and for Ritchey–Chrétien telescope primaries. I am particularly impressed by one very practical feature of Charles's designs: he insists on using spherical surfaces because of the ease of manufacture and testing that this allows. This is not just because the glassmakers tell him so; he knows it from direct experience, dating back to wartime optical work.

Some would say that optical design is now easy, because of the vast computer power which can now be applied to optimisation. It might even appear that the computer can be told to use, say, three slabs of plane glass, and to bend their surfaces and move them around until the required optimisation is complete. This simply does not work. The designer has to understand conventional aberration theory as well as knowing what types of glass are actually available in a suitable quantity. But above all he has to start with experience and flair, starting the computer program only when there is a good chance that the optimisation will be worth making.

The UK telescopes have benefited from all this in many ways. I would pick out particularly the correction of atmospheric dispersion, the Low Dispersion Spectrograph, and the outstandingly successful Faint Object Spectrograph, the FOS. The FOS shows us that optical design need not become more and more complex; it just needs an appreciation of the whole problem, including the new detector systems, and an essential element of flair. This is what Charles has provided for us, and we shall always be in his debt for his doing so.

II Current instrumentation, development, results

2
Ten Years of the Manchester Echelle Spectrometers

John Meaburn* and Myfanwy Bryce*

Abstract

Two versions of a simple echelle spectrometer have been used on the Anglo-Australian, Isaac Newton and William Herschel Telescopes. The primary purpose of this design is the investigation of extended, astronomical, emission-line sources in the wavelength domain 3900–7000 Å. A high optical efficiency is achieved as a consequence of this degree of dedication. Many easily interchangeable options for the entrance area of the spectrometer have increased its flexibility during observations.

1. Introduction

The initial version of the Manchester Echelle Spectrometer (MES) was manufactured in 1982 and used for the first time on the Anglo-Australian Telescope in 1983 [Meaburn *et al.*, 1984]. A later version was commissioned on the Isaac Newton Telescope in 1986 and William Herschel Telescope in 1987. The aim was to take advantage of some degree of dedication in the design of these spectrometers; their primary use is to obtain spatially-resolved profiles of individual emission lines from extended sources emitting within the limited wavelength range of 3900–7000 Å. For this purpose a spectral resolution of $R \leq 10^5$ is required. For such tasks they remain, for a variety of reasons, highly competitive spectrometers. However, in practice they have had much wider application within this wavelength domain simply because they were available on the telescopes with the required spectral resolution.

Here, the principles that should guide the design of echelle spectrometers for work on faint, extended, emission-line sources at high spectral resolution will be summarised. The manifestation of these principles in the form of MES will be shown. The importance of having a variety of options for the 'input' area, other than simply a long slit, will be emphasised. Ways of accurately calibrating echelle spectra will be considered.

*Department of Astronomy, University of Manchester, Manchester M13 9PL, UK.

9

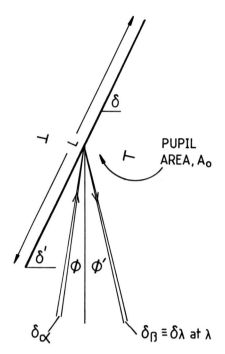

Figure 1. The essential parameters of an echelle grating operating in a collimated light beam are shown. Note that the grating angle δ' can be different from the blaze angle δ. L is the length of the grating filled by the beam.

2.   Design principles

Why use an echelle grating for this primary astrophysical objective? The reasons are complex and, as with most instrumental solutions, there are viable alternatives not employing echelle gratings. However, first consider that for any spectrometer working on faint, extensive sources the product

$$\text{merit} = \epsilon \times R \times \delta\alpha \times A \times l \tag{1}$$

should be maximised to obtain the greatest number of photons in a fixed integration time in each spatial and spectral element, where ϵ is the optical efficiency of the whole system (including the telescope), R is the spectral resolution ($\lambda/\delta\lambda$) as determined by the angular slit width $\delta\alpha$ to the grating, A is the (exploited) pupil area on the grating and l the angular slit length. For a given pupil area, A_0 (see Fig. 1), $A = A_0(1 - \tan \delta \tan \phi)$ due to groove masking when $\phi = \phi'$ and $\delta = \delta'$. Therefore A decreases at a fixed δ for increasing ϕ.

Then consider the diffraction relationships for the grating with blaze angle δ shown in Fig. 1 within the Ebert configuration (where the grating angle $\delta' = \delta$,

Table 1. The performance of reflection gratings is illustrated where $\delta = \delta'$ and $\phi = \phi'$ in Fig 1. The fractional area loss is for groove masking which diminishes A in Eqn 1 for a fixed pupil area A_0.

δ	ϕ	$R\,\delta\alpha$	Fractional area loss	$R \times \delta\alpha \times A$	$R\,\delta\beta$
64.5°	0°	4.2	0.0	4.2 A_0	4.2
	20°	17.7	0.76	4.2 A_0	2.4
20°	0°	0.73	0.0	0.73 A_0	0.73
	20°	0.84	0.14	0.73 A_0	0.64

$\phi = \phi' > 0°$ and $\delta\beta$ is the width of the resolution element $\delta\lambda$ at λ), i.e.

$$R\,\delta\alpha = \frac{2\sin\delta\cos\phi}{\cos(\delta + \phi)} \qquad (2)$$

and

$$R\,\delta\beta = \frac{2\sin\delta\cos\phi}{\cos(\delta - \phi)}. \qquad (3)$$

Note that (i) for increasing ϕ in Eqns 2 and 3, $R\,\delta\alpha$ increases and $R\,\delta\beta$ decreases and (ii) for the quasi-Littrow configuration with $\phi \approx 0°$ inserted in Eqns 2 and 3 then

$$R\,\delta\alpha \approx R\,\delta\beta \approx 2\tan\delta. \qquad (4)$$

It is illustrated in Table 1 for a fixed δ that the increase in $R\delta\alpha$ as ϕ increases (Eqn 2) is exactly offset through groove masking by the reduction in A for a fixed pupil area A_0; the value in Eqn 1 of $R \times \delta\alpha \times A = 2\tan\delta\,A_0$ is an increasing function of δ and independent of ϕ (see Table 1).

The quasi-Littrow configuration with an echelle grating with δ large is then an optically easy way of (i) obtaining $R \leq 10^5$ for a fixed $\delta\beta$ (see Eqns 3 and 4) and (ii) permitting l to be large. In this configuration, with $\delta' = \delta$ and $\phi \approx 0°$, $R \times \delta\alpha \times A$ is then maximised in Eqn 1 by using the largest practical value of δ (see Table 1).

3.   Practical parameters

Bausch & Lomb echelle gratings with $\delta = 63.43°$ (nominally – see later) and $\approx 76°$ are available. The difficulties of using the $\delta = 76°$ grating compared with the 63.43° one can be appreciated in Fig. 1. The grating length, L, that is required to accept all of the fixed beamwidth becomes excessive for the $\delta = 76°$ case. For MES the 31.6 grooves/mm Bausch & Lomb grating with δ nominally 63.43° was chosen; for the 76° case one would have to be mosaiced to achieve the desired length, L = 41 cm, and the further factor two in merit (Eqns 1 and 4). Incidentally, Diego [Diego, 1987] has reported that $\delta = 64.57°$ for this 31.6 grooves/mm B & L grating.

11

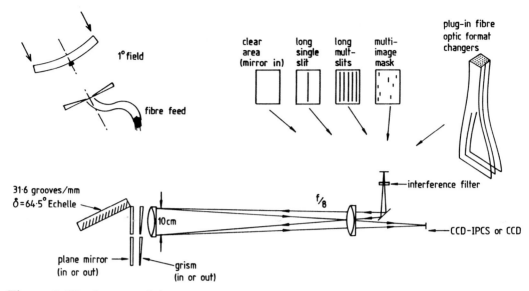

Figure 2. The layout of the Manchester Echelle Spectrometer is shown with various options for the entrance area. A plane mirror and grism can be inserted in the collimated beam either separately (for imaging) or together (for low dispersion spectrometry). MES can also be fibre-optically coupled to a 20-cm aperture telescope.

An independent measurement in MES has confirmed his result to within $\pm 0.1°$ by unambiguously identifying the echelle orders for a wide range of lamp lines.

The quasi-Littrow layout of MES is shown in Fig. 2. A dioptric solution for the collimator/camera was chosen for various reasons, not least that Charles Wynne came up with a two contact-doublet lens solution which has only four air/glass surfaces [Meaburn et al., 1984]. The efficiency ϵ is maximised by coating these surfaces with 3-period anti-reflection (ar) coatings with 0.5% reflectivity per surface in the working wavelength range. For longer wavelengths this reverts back to 4%. The flint glass in the lenses absorbs below 3700 Å. The slits are chromium-deposited and sandwiched between two glass plates with similarly ar-coated outer surfaces. The only regret is that a plane mirror was chosen to divert the incoming beam to the grating. An ar-coated prism with a sealed, aluminised, internally-reflecting, hypotenuse surface would be preferable (and can still be fitted). The prism/collimator/camera/sealed-slit element of this all-dioptric solution has $\epsilon = 94\%$ and needs little maintenance over many years when compared with its reflecting alternative. The limited budget, and use at the more efficient and flexible Cassegrain focus, constrained A for a beam diameter of 10 cm. Without restrictions this could be 25 cm with a single 64.5° echelle grating, though increasingly impractical at this larger diameter for a 76° echelle ($L = 103$ cm; see Fig. 1). In the present design, individual echelle orders are isolated by efficient 3-period interference filters

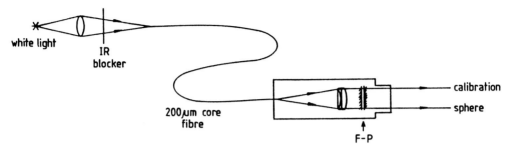

Figure 3. The layout of the Edser–Butler calibrator is shown. The gas-spaced Fabry–Perot is in a container with the same dimensions as a standard calibration lamp.

(square bandpass) whose outside surfaces are again ar coated. No cross dispersion is included, though a grism and plane mirror can be inserted to give a quick-look low-dispersion spectrum of any source. With the plane mirror alone inserted, a filtered image of the field can be obtained and, if required, the exact position of the slit recorded. The favoured detectors have been the IPCS [Boksenberg & Burgess, 1973] and the CCD-IPCS [Fordham *et al.*, 1986] but these are now giving way, particularly in the red, to the large CCDs that are becoming available.

4.   Spectral calibration

In principle, lines from an emission-line lamp in a known echelle order, n, isolated by an interference filter, can be used to calibrate long-slit spectra obtained through another filter isolating any other echelle order, n'. Any lamp line of wavelength λ_L appears in n' to have wavelength

$$\lambda' = \lambda_L n/n'. \qquad (5)$$

For reasons not yet understood Eqn 4 only applies in MES to the required ± 1 km/s accuracy when n' is within ± 15 orders of n. Apart from this limitation it is often difficult to find lamp lines spread evenly over the working order. Consequently two methods are under development to put evenly-spaced calibration profiles over the whole detector format of MES.

4.1.   The Edser–Butler calibrator

A potentially improved way of spectrally calibrating echelle spectra employs the Edser–Butler (EB) Fabry–Perot shown in Fig. 3. This device, when fed with on-axis white light by the fibre, produces a set of long-slit fringes evenly spaced in wave-number over the whole detector format which contains one echelle order in MES. See [Meaburn *et al.*, 1992] for a full technical description and Fig. 4 for a plot of the EB fringes from the centre section of a long-slit MES spectrum used to calibrate

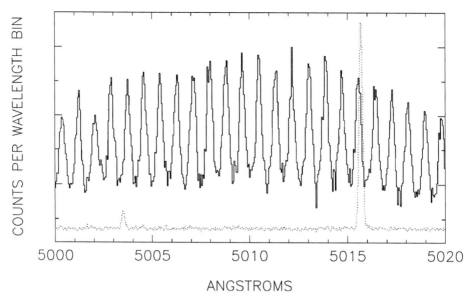

Figure 4. Edser–Butler fringes in the 114th order of MES are compared with the profiles of two lamp lines. The brightest is HeI 5016 Å.

the 114th echelle order which contains the [OIII]5007 Å nebular line. Here, the fringes are separated by 0.843 Å and have halfwidths of ≈ 0.14 Å. Their absolute wavelength can be identified to ± 0.02 Å by employing the method of exact fractions to identify unambiguously any Fabry–Perot order.

The optically-contacted etalon is air-spaced (refractive index μ) to minimise the thermal drift $\delta \lambda$ of the fringes for

$$\delta \lambda / \lambda = \delta \mu / \mu = 10^{-6} \delta T \qquad (6)$$

for air with a temperature change δT K. For the solid, and more mechanically stable, alternative, made from fused silica, this ratio is as high as $6.6 \times 10^{-6} \delta T$ [Georgelin, 1970]. The mechanical stability of the gas-spaced etalon was measured by swinging MES on the Isaac Newton Telescope through $60°$ and comparing the fringe shift with respect to the lamp line seen in Fig. 6. No shift of $\delta \lambda / \lambda \geq 3 \times 10^{-6} (\equiv \pm 1$ km/s) was found, which is more than adequate stability for the present purposes. This calibrator will also be most useful for cross-dispersed echelle spectra over large wavelength ranges. In this case, careful consideration would have to be taken of the wavelength dependence of the phase changes on reflection of the thin silver (or aluminium) reflecting layers (see [Bennett, 1964]).

14

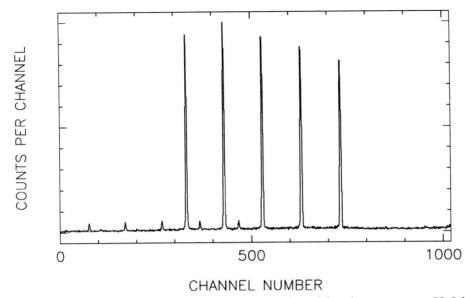

Figure 5. Five-element, multi-slit spectra produced by the same two HeI lamp lines shown in Fig. 4.

4.2.   Self calibration with a multi-slit

A multi-slit in the entrance area (see Fig. 2) when illuminated by one or two lamp lines can be used to provide a grid of calibration profiles in one echelle order. This is illustrated in Fig. 5 where a cut through the centre section of a spectrum obtained by illuminating a 5-element multi-slit with two HeI lamp lines is shown. To use a multi-slit with $(2 \mid m \mid + 1)$ separate slits (e.g. $m = 2, 1, 0, -1$ and -2 respectively for a five-element one with the central slit denoted $m = 0$), for the purposes of calibration, the equivelent wavelength λ_{out} for an illuminating line of wavelength λ_L must be calculated. Assume the slits are separated by δx mm and the collimator has a focal length f mm; then the input angle (see Fig. 1) of the mth slit is given by

$$\phi = \phi_o + (m \, \delta x 180 / f \pi) \tag{7}$$

where ϕ_o is the input angle of the central slit ($m = 0$). In this case, for a groove separation a and echelle order n, the output angle of the corresponding mth slit is

$$\phi' = \delta' - \sin^{-1}(n \lambda_L / a - \sin(\phi + \delta')) \tag{8}$$

to give

$$\lambda_{out} = (a/n)(\sin(\delta' + \phi) + \sin(\delta' - \phi')). \tag{9}$$

The drawback of this method is that δ' and ϕ have to be known in Eqn 8 to $\pm 0.05°$ to achieve the ± 1 km/s calibration accuracy. However, these values can be

15

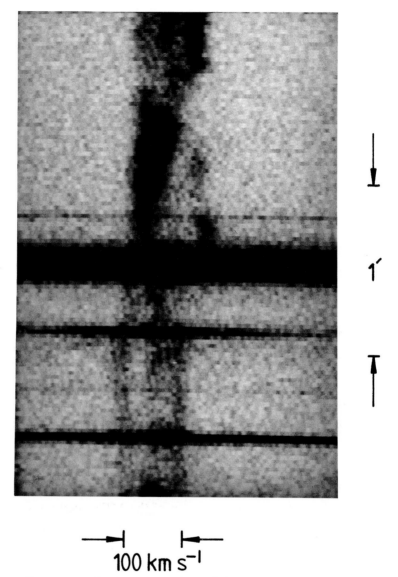

Figure 6. A negative greyscale representation of the MES position–velocity array of Hα profiles over the faint core of the LMC giant shell N11. The dark horizontal bands are stellar spectra. This was obtained with MES on the Anglo-Australian Telescope with a 10 km/s resolution.

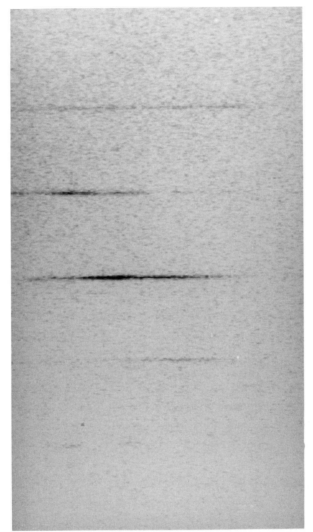

Figure 7. A negative greyscale representation of the MES position–velocity array of [OIII]5007 Å profiles from the faint halo of the planetary nebula NGC 6543. This was obtained with MES on the Isaac Newton Telescope with a 3 km/s resolution.

determined by illuminating the central slit of the multi-slit with a rich lamp spectrum and then all slits with the single lamp line. The measured equivalent wavelengths can be used for all subsequent multi-slit self calibrations.

5. Astronomical applications

The various options for the entrance area of MES are shown in Fig. 2. The single long slits, long multi-slits, multi-image masks and the clear area are on a motor-

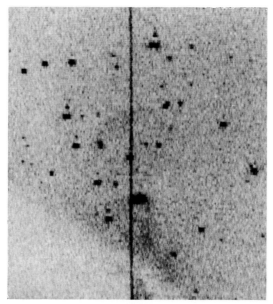

Figure 8. The slit position is recorded against a quick-look image of the field. Both exposures were obtained with the plane mirror inserted (see Fig. 2).

driven slide and can be quickly (5 sec) interchanged while observing. Similarly the plane mirror can be 'flicked' in and out of the collimated beam. The grism for low-dispersion spectroscopy and the compact fibre-optic format changers are loaded manually. Examples of astronomical data obtained with some of these options will now be presented.

5.1.   Long, single and multi-slits

A grey-scale representaion of the position-velocity array of Hα profiles over the faint central 'hole' of the giant shell N11 in the Large Magellanic Cloud is shown in Fig. 6. The width of the single slit was $150\,\mu$m ($\equiv$ 10 km/s) and MES was combined with the Anglo-Australian Telescope for this observation. Many extensive velocity components are apparent which it is argued [Meaburn et al., 1989] are caused by successive supernova explosions.

The use of a 5-element multi-slit, in MES on the Isaac Newton Telescope with slit widths of $30\,\mu$m ($\equiv$ 3 km/s) is demonstrated, similarly, in Fig. 7. The [OIII]5007 Å profiles from the giant, faint, halo of the planetary nebula NGC 6543 are just resolved demonstrating that this kinematically-inert gas must originate in the slow, dense wind of the red giant precursor [Bryce et al., 1992].

The use of a device to record precisely the position on the sky of the long, single or multi-slits is shown in Fig. 8. Here, after a spectral observation, the plane

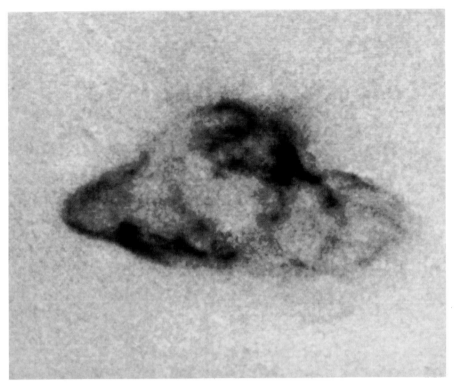

Figure 9. A negative image in the light of Hα of the peculiar supernova remnant CTB 80, obtained with MES in its imaging mode on the William Herschel Telescope.

mirror (see Fig. 2) is flicked in and a sky-illuminated image of the slit recorded. The clear area is then driven in to replace the slit to give a quick image of the field superimposed on the slit image.

5.2.   Imagery

Direct filter imagery with the plane mirror and the clear area both inserted (see Fig. 2) is illustrated in Fig. 9 where a continuum-subtracted Hα image of the peculiar supernova remnant CT80 is shown. This was obtained with MES combined with the William Herschel Telescope. A pulsar, slung out from the explosion site, is ploughing through the remnant shell to cause the ionisation that is revealed [Whitehead *et al.*, 1989].

'Velocity imagery' is illustrated in Fig. 10. A data cube with two spatial and one velocity dimension has been obtained with MES on the Isaac Newton Telescope, in the light of the [OIII]5007 Å line, by obtaining four integrations over the Herbig Haro-like object M16 HH1. Between each integration the telescope has been stepped in a direction perpendicular to the slit length by 0.2 of the slit

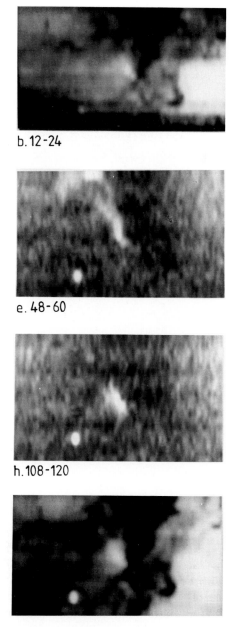

b. 12-24

e. 48-60

h. 108-120

j. all velocities.

Figure 10. Positive images, in the light of [OIII]5007 Å, of the Herbig–Haro-like object M16 HH1 in various ranges of radial velocity (marked below each frame in units of km/s). These were obtained with MES on the Isaac Newton Telescope, from a data cube produced by stepping a five-element multi-slit across the source.

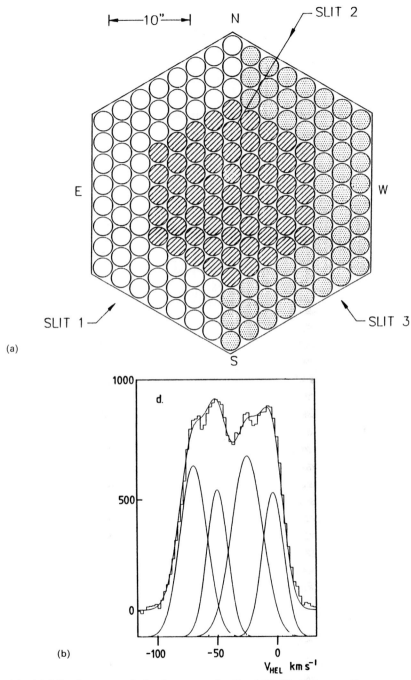

Figure 11. (a) The layout of the input end of a 169-fibre array. Separate areas of the array feed three entrance slits, 1–3, of MES simultaneously. The circles are for the 490-μm plastic buffers of the fibres which have 200-μm cores. (b)An [OIII]5007 Å profile from one of the fibres in Fig. 11(a) illuminated by the Dumb-bell nebula, obtained with MES on the William Herschel Telescope. A four-Gaussian fit is shown.

21

separation as projected on the sky. Images in different velocity intervals can then be displayed. Extensive outflows of highly-ionised gas appear to be associated with this lowly-ionised knot in which a young stellar object is likely to be embedded [Meaburn & Whitehead, 1990].

Another form of imagery (not illustrated) has proved most useful with MES in very special circumstances. Here, a very broad slit whose width is $\equiv 10\,\text{Å}$ is inserted and an echelle spectrum obtained in say the Hα and [NII]6584 Å lines simultaneously. If the source is emitting very narrow lines, similar to that shown in Fig. 7, images in each of the lines and neighbouring continuum are formed in each integration in which the angular resolution is preserved in the direction of dispersion. Very accurate line-ratios can be obtained over the area of the source as well as very good continuum subtraction.

5.3. Fibres

The use of a 20 cm-long, compact, fibre-optic format changer is illustrated in Fig. 11a & b. The 169 fibres, in the hexagonal format, shown in Fig. 11a feed three long, parallel, entrance slits of MES on the William Herschel Telescope ([Meaburn et al., 1992] describe details of this development). The aim is to obtain line-profiles from sources whose geometry does not match that of long, thin, entrance multi-slits. An example of an [OIII]5007 Å profile from the core of the Dumb-bell nebula is shown in Fig. 11b. This is from one of the 169 slits in the array and shows four separate velocity components indicating the presence of two highly-ionised, expanding shells in the nebular core.

A further development is complete. This is the fibre-optical coupling of the output of a 20 cm aperture, Meade, telescope to three parallel, long, entrance slits of MES (see Fig. 2). Each fibre core projects as 20″ on the sky. The 169 fibres are distributed in a 1.0° diameter array in the focal plane of the telescope. The intention is to obtain line profiles from faint phenomena extending many degrees over the sky.

5.4. Low dispersion

A quick-look, low-dispersion mode is incorporated. Here both the plane mirror and a grism are inserted simultaneously into the beam (see Fig. 2). A 79 Å/mm second-order spectrum, from 4300 to 6800 Å of the source on the same entrance slit as used for an echelle spectrum can be obtained. Examples of echelle and grism spectra of the symbiotic star CH CYG, obtained in this way with MES on the Isaac Newton Telescope, are shown in [Bode et al., 1991].

Incidentally the dispersed spectrum of a grism scarcely moves as the grism is rotated in the plane of the dispersion. Consequently to move the 79 Å/mm spectrum across the detector, the plane mirror has to be tilted. However, the grism itself has to be tilted to get rid of a complex array of higher-order ghost spectra. These are produced by internal reflections within the grism; therefore, as for the spectrum of a reflection grating, they move off the detector with rotation of the grism in its dispersion plane. Ideally the grism should be tilted a few degrees perpendicular to its dispersion plane for this purpose. In this case the blaze peak will be maintained at the prescribed wavelength.

References

[Bennett, 1964] Bennett J.M., *J. Opt. Soc. Am., Vol. 54 (1964), p. 618.*

[Bode *et al.*, 1991] Bode M.F., Roberts J.A., Ivison R.J., Meaburn J., Skopal, A., *Mon. Not. R. astr. Soc., Vol. 253 (1991), p. 80.*

[Boksenberg & Burgess, 1973] Boksenberg A., Burgess D.E., *Proc. Symp. TV Sensors, eds Glaspey J.W. & Walker G.A.H., University of British Columbia: Vancouver, 1973, p. 21.*

[Bryce *et al.*, 1992] Bryce M., Meaburn J., Walsh J.R., Clegg R.S.E., *Mon. Not. R. astr. Soc., Vol. 254 (1992), p. 477.*

[Diego, 1987] Diego F., *Appl. Optics, Vol. 26 (1987), p. 4714.*

[Fordham *et al.*, 1986] Fordham J.L.A., Bone D.A., Jorden A.R., *Proc. Soc. Photoopt. Instr. Eng., Vol. 627 (1986), p. 206.*

[Georgelin, 1970] Georgelin Y.P., *Astr. Astrophys., Vol. 9 (1970), p. 436.*

[Meaburn *et al.*, 1984] Meaburn J., Blundell B., Carling R., Gregory D.F., Keir D., Wynne, C., *Mon. Not. R. astr. Soc., Vol. 210 (1984), p. 463.*

[Meaburn *et al.*, 1989] Meaburn J., Solomos N., Laspias V., Goudis C., *Astr. Astrophys., Vol. 225 (1989), p. 497.*

[Meaburn & Whitehead, 1990] Meaburn J., Whitehead M.J., *Astron. Astrophys., Vol. 235 (1990), p. 395.*

[Meaburn *et al.*, 1992] Meaburn J., Christopoulou P.E., Goudis C.D., *Mon. Not. R. astr. Soc., Vol. 256 (1992), p. 97.*

[Whitehead *et al.*, 1989] Whitehead M.J., Meaburn J., Clayton C.A., *Mon. Not. R. astr. Soc., Vol. 237 (1989), p. 1109.*

3

Developments in Optical Systems for Infrared Astronomical Instruments

T.J. Lee, with J.W. Harris, E. Atad and C.M. Humphries *

Abstract

Factors influencing the design and development of optical systems for infrared instruments are described and some of the systems produced at the Royal Observatory are presented. These span the early days of UK infrared astronomy with smaller telescopes and simple detectors to present-day complex instruments with two-dimensional arrays of detector which are analogues of cameras and spectrographs with which visible-wavelength astronomers are more familiar. Infrared instruments have constraints over and above those for visible wavelengths because of the requirement to cool the entire optics; however, a wider range of material properties provides interesting choices (though this can introduce new problems), while the longer wavelength relaxes surface specifications just enough to permit novel manufacturing techniques to deliver required performance. Further development of some of the methods used in infrared instruments may be of benefit to shorter-wavelength applications.

1. Introduction

Since the early 1970s UK astronomers have had access to telescopes dedicated to infrared use: first the 1.5 m Flux Collector in Tenerife[1], and since 1980 the 3.8 m UK Infrared Telescope (UKIRT) in Hawaii. The difference in aperture, focal ratio and image quality posed different constraints on optical systems for infrared instruments which used single detector-elements until the mid-1980s. The availability of two-dimension pixel devices for the infrared again led to changes in design approach for instruments such as intermediate-dispersion spectrographs for the infrared. These developments are traced, comparing and contrasting with the practices of visible-wavelength astronomy.

---

*Royal Observatory, Blackford Hill, Edinburgh EH9 3HJ, UK.
[1]Now known as the Carlos Sánchez Magro Telescope of the Instituto de Astrofísica de Canarias, La Laguna, Tenerife – *Ed.*

24

2.   Basics

Fundamental factors which influence the performance of optical systems for all wavelengths are:

Diffraction

Rayleigh criterion $\alpha = 1.22\lambda/D$

Limits image size

Etendue

$A\,\Omega$ is conserved

or $D\theta < d\phi$

Limits field size

Practical factors which further concern us are:

Atmosphere

Material properties

Manufacturing limits on optical elements and systems

Detectors

For example, a 2 m telescope has a diffraction limit of 0.06 arcsec at 0.5 μm which is well within atmospheric seeing limits. Diffraction can be disregarded. But a 2 m telescope has a diffraction limit of 1.2 arcsec at an infrared wavelength of 10 μm, clearly larger than seeing limits.

Etendue considerations mean, for example, that if we have a detector 1 cm across and a final image focal ratio of f/1 then for a 4 m telescope the maximum field size on a 1 cm detector is 8.5 arcmin. For a 10 μm pixel it would be 0.5 arcsec. We know from visible optics that it is difficult to build systems with a relative aperture approaching one without serious aberrations in the field, e.g. 35 mm cameras, astronomical instruments. It is not possible to image an arbitrarily large field on a detector of given size. Optical design becomes difficult for $f/\# > 2$.

3.   Early infrared instruments

The first systems we built in Edinburgh were for the 1.5 m flux collector in Tenerife. How demanding were they?

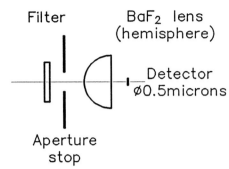

Figure 1. The Fabry lens system for the stellar photometer on the 1.5 m Tenerife Flux Collector.

Instrument: stellar photometer

Image size (arcsec)	Aperture (mm)	Telescope aperture (arcsec)	Instrument detector size (mm)	Relative aperture
2	1500	15	0.5	4.6

General texts for designers such as *The Infrared Handbook* [Wolfe & Zissis, 1985] can be used as a guide for optical as well as other aspects of design. A Fabry lens system was chosen for the instrument, shown in Fig. 1. The primary mirror is imaged onto the detector which acts as a stop, i.e. we image so that all rays passing through the primary are incident on the detector. A focal-plane aperture selects the field of view.

We want the blur β due to this lens to be less than that due to telescope image quality of 2 arcsec. This requirement works out as $\beta < 2 \times 10^{-3}$ radians.

As a starting point we use the blur chart for spherical aberration from *The Infrared Handbook* [Wolfe & Zissis, 1985]. Such a chart graphs blur-spot angular size as a function of numerical aperture and refractive index. This tells us that for

refractive index $= 1.5$, and

numerical aperture $f/\# = 4$,

we get blur $= 10^{-3}$. Therefore we can achieve our goal for $f/\# > 4$ if we choose the shape of lens which minimises spherical aberration. Graphs of solutions to the appropriate equations are published [Wolfe & Zissis, 1985] and indicate the appropriate lens shapes for different refractive indices. Barium fluoride, a material transparent in the infrared with $n \approx 1.5$ can be used. Thus for the 1.5 m flux collector instruments, adequate design could be carried out using simple lenses and the aid of published charts.

When we started building instruments for UKIRT it became immediately obvious that design would become more difficult because the telescope is larger by a factor of 2.5 and the image quality specification was better by a factor of at least two. Effectively for the same instrument the etendue consideration meant a final numerical aperture 2.5 times smaller, i.e. faster than f/2. Inspection of the blur charts showed us that the blur becomes greater by more than an order of magnitude with the same optical design.

The actual specification placed on the photometer performance was more demanding since capability for spectrophotometry and for nebular photometry implying a larger aperture were added. The relevant parameters are as follows:

Instrument: spectrophotometer

Image size (arcsec)	Aperture (mm)	Telescope aperture (arcsec)	Instrument detector size (mm)	Relative aperture
1	3800	20	0.5	1.4

Because this specification starts to approach the theoretical limits for simple systems it became necessary to develop a design strategy and to ray-trace the designs.

The approach to design was signposted through the analytical formulae for aberrations:

$$Spherical\ \beta = \frac{n(4n-1)}{128(n+2)(n-1)^2(f/\#)^3}$$

$$Coma\ \beta = \frac{\theta}{16(n+2)(f/\#)^2}$$

These tell us to use materials with greater refractive index n to reduce the blur β. Alternatively one can use more elements in a group of low-index lenses. In the infrared there is a choice of materials with refractive indices up to $n = 4$.

The adopted solution is shown in Fig. 2; it is to use a higher-index material. Here the telescope secondary mirror is imaged using a single Zinc Seleride (ZnSe; $n = 2.5$) element with the "bend" chosen to minimise spherical aberration. A stop is placed at this image to eliminate rays from the surroundings. This stop is cold, < 80 K. Such stops are very important in infrared instruments. They limit the amount of warm sky (or worse, mechanical structure) seen by the detector which gives rise to noise, background photons. A good image at this stop is essential. This 1.5-mm diameter image is relayed onto the 0.5-mm diameter detector using a spherical lens, again of ZnSe. Why use a sphere? Essentially the numerical aperture target of 1.4 demands large curvature surfaces, and feasibility of manufacture led to

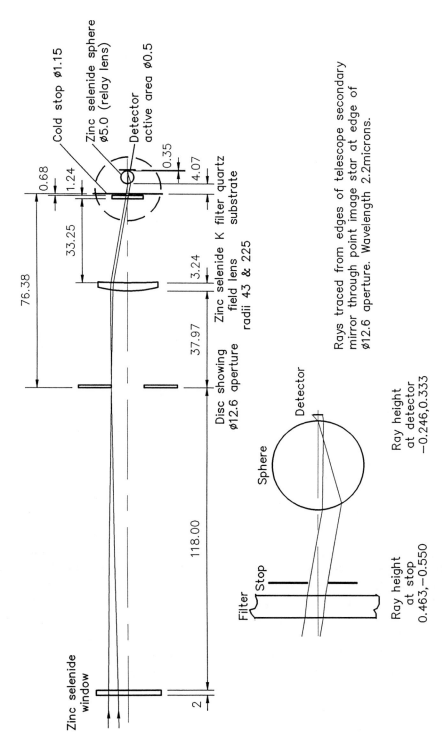

Figure 2: The optical system of the initial UKIRT spectrophotometer.

the choice of a sphere. This is not the best shape from the viewpoint of spherical aberration; optimal shape would have given a factor of two less spherical aberration.

Performance achieved in practice was an 18 arcsec beamwidth as opposed to the 20 arcsec requested. This was judged to be acceptable.

4. Difficulties of IR optics

There are of course penalties in using a high-refractive-index material. Reflections at the vacuum–lens interface are high:

$$\rho = \left(\frac{n-1}{n+1}\right)^2$$

– for $n = 2.5$ (ZnSe) it is 18% per surface.

As at shorter wavelengths, anti-reflection coatings reduce this loss but in this case it remains high. Best values are 2% for narrow bands but for 1 to 5 μm an average loss of 5% is the best obtained.

In practice we found that a problem with reflections more severe than this loss is "ghosting" or the detection of photons other than at their first pass. In this particular design the back surface of the lens acts as a collimating mirror which when combined with the interference filter acting as a mirror causes additional rays in the centre of the focal-plane aperture to fall on the detector. The centre-to-edge response for a point source might vary by 3% for the aperture size typically used for point source photometry.

Though modern ray-tracing packages now do include ghost analysis, this may not in fact be adequate for infrared systems. This is because rays are extinguished on their third reflection on the assumption that the reflectivity of surfaces in optical systems is small.

Sensitive infrared detectors need to be cooled to cryogenic temperatures for operation. Optics for infrared systems have to be cooled also. Differential expansion of optical components and supports would, in principle, cause defocus or misalignment. Since cold optics must operate in a vacuum, re-adjustment is not normally possible. Therefore it is essential to assemble optics in such a way that on cooling they move into adjustment. This requires care in design and a knowledge of the properties of each optical component. The same factors influence tolerancing in infrared optics as in visible-wavelength optics. Because the same (or better) image quality is demanded by astronomers, the problem is no easier.

For this spectrophotometer an adequate design could be produced using simple ray-tracing provided advanced materials were used. High-refractive-index materials must be anti-reflection coated, and ghost analysis is essential.

5. Recent systems for large telescopes

The first two systems we discussed used single element detectors. About five years ago infrared array detectors became available to astronomers. Several observatories now own infrared cameras similar to CCD cameras. The main difference is that the complete camera including filters and optics is cooled, not just the detector. This advent of area detectors quickly produced a demand for spectrographs analogous to those used at optical wavelengths. In the UK the UKIRT Users Committee asked for a long-slit spectrograph with intermediate resolution (300 to 40000) resolving power. Described briefly here, the instrument CGS4 is reported on more fully by [Mountain *et al.*, 1990].

5.1. CGS4, a cryogenic Intermediate Dispersion Spectrograph: goals

R = 300-2000	R = 10000-30000
Line identification	
Total flux measurements	Gas kinetics of galactic phenomena
Broad feature spectroscopy	
Extragalactic	Atomic and Molecular Physics

Long slit	> 90 arcsec
Range	1 to 5 μm
High sensitivity –	Slit width, pixel match
Image quality –	80% of light onto 30 μm pixel
High throughput	> 25

Operability:
    Stable
    Low heat input
    Interfaces
    Reduction software

5.2. Constraints

The wavelength range of 1 to 5 μm requires a detector temperature of 30 K and optics temperature of 70 K. These cryogenic requirements imply a compact envelope to reduce heat input (surface area). Resolving power requires a large grating, 270 × 150 mm which opposes a small size of instrument. The long slit and wide spectrum require a finite field and a flat one; this challenges optical design. The trend for large-format detectors is to 30 μm size pixels. Etendue considerations require a final $f/\# = 1.35$ to meet the pixel size. Mechanical tolerances are tight – as tight as visible instruments, yet the structure must be lightweight (cryogenics). High throughput requires minimum losses through reflection or vignetting. There is no access-to-instrument when it is operating in the infrared.

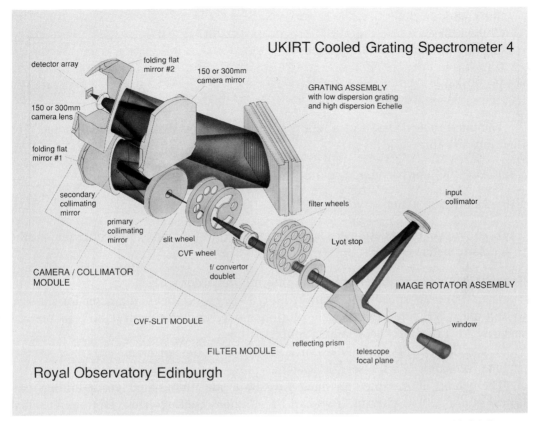

Figure 3. The design for UKIRT Cooled Grating Spectrometer 4 (CGS4).

5.3. Solutions – gold and aluminium

Clean pure gold surfaces have reflectance of about 99% at infrared wavelengths. A few extra reflections are worth trading for an advantage of some kind.

Aluminium alloys can be
Cast
Welded
Machined
Diamond-machined to 1/4 wavelength (visible)

The use of gold coatings and aluminium structures and optics is the key to successful design implementation.

The optics of CGS has the following eight stages, and Fig. 3 shows how these are set out:

Input optics
Filtering

31

f-converter
Slit area
Collimator
Grating mount
Camera
Focal plane

The *input optics* consist of a reflecting prism and a concave spherical image. The functions are three-fold:

(a) As a "K" mirror image-rotator

(b) As an area which can be baffled for stray light

(c) To image the telescope secondary to permit a cold Lyot Stop (as in the photometer)

The *filtering stage* contains broad pre-filtering for order sorting.

The *f-converter* speeds the beam from f/35 to reduce beam diameter for slit size, to suit the CVF filter for high resolution mode, and the central hole in pierced mirrors. An LiF/BaF_2 achromatic doublet is used.

A selection of slits is mounted on a slit wheel in the *slit area*.

The *collimator* is a Cassegrain system.

The *grating* is mounted in quasi-Littrow mode illuminated via a folding flat. Gratings are 270×150 mm. Two can be mounted back to back. Options installed are 75 l/mm, 150 l/mm, or the 31.6 l/mm echelle.

The *camera* also images via a folding flat. Two cameras are available: one of 150 mm focal length using an aspheric mirror and a ZnSe convex lens with spherical surfaces to yield a final numerical aperture of f/1.35; and the other of 300 mm focal length, an f/2.7 camera which is an aspheric mirror with a BaF_2 and LiF doublet.

5.4.   Optimisation of the optics

Even for a small field of 90 arcsec and 10 mm of spectrum, an f/1.35 design is a challenge. Our trial design of a classical Cassegrain collimator and a conic aspheric camera did not fully meet specification. We thus optimised the design using the Code V ray-tracing software.

Constraints were as follows:

(a) The lenses must have spherical surfaces for ease and cost of manufacture.

(b) The collimator must remain Cassegrain (parabolic/hyperbolic) but the conic constants can be changed (non-classical); the beams onto the grating may not be quite parallel.

(c) The camera mirror can be general aspheric.

Code V can ray-trace the model with the actual diffraction grating in place rather than a mirror in its stead.

Using this software we were able to obtain point-spread-functions of 80% EED within 30 μm over the 3 mm $\times$ 10 mm focal plane for both f/1.35 and f/2.7 cameras, this method of optimisation minimising the number of surfaces required to achieve a given optical performance.

5.5.   Tolerancing the optics

Code V has features for calculating the effect of surface quality and errors of positioning of optics on image quality. By choosing to mount the higher-powered components together into a module we were able to specify a decentre tolerance of 25 μm and a tilt of 0.3 millirad all of these components. A decentre tolerance of 100 μm was allowed on components not mounted in the camera/collimator module. Axial-distance tolerances were 50 or 100 μm. Most surfaces have a half-fringe irregularity specification. These tolerances allow the image-quality specification to be met.

Only one adjustment is permitted, an axial movement of the convex secondary of the collimator. The detector focal plane can be moved in focus to compensate for chromatic aberration in the f/1.35 camera lens. This is significant for $\lambda < 1.6\,\mu$m.

5.6.   Manufacture of optics

Except for the BaF_2 and LiF components all the optical components were made by diamond-point cutting.

As the mirrors and ZnSe lens are machined components, they were manufactured to optimally fit the optical footprints, thus reducing their thermal mass. Reference diameters and alignment surfaces were diamond-machined in the same operation as that for the optical surfaces. This way the optical surfaces can be referenced to the optical structure with almost optical precision. The aluminium substrates were manufactured at ROE in 6061-T6 and thermally cycled between machining operations as described by [Erikson et al., 1984]. The diamond machining (turning and fly-cutting) was done by Diamond Electro-Optics. The gratings were replicated onto diamond-machined aluminium blanks by Milton Roy.   An example of one of the more complex components is the camera mirror shown in Fig. 4. It has a general aspheric form given by the polynomial shown. To test this component after machining, a computer-generated hologram (CGH) was used as described by [Smith, 1981].   The interferogram (insert) shows the difference between the holographic representation of the true surface (generated by Code V) and the actual diamond-cut surface. On first inspection it was clear that the central part of the mirror was outside our specification of 1 fringe at 0.659 μm. However, when the

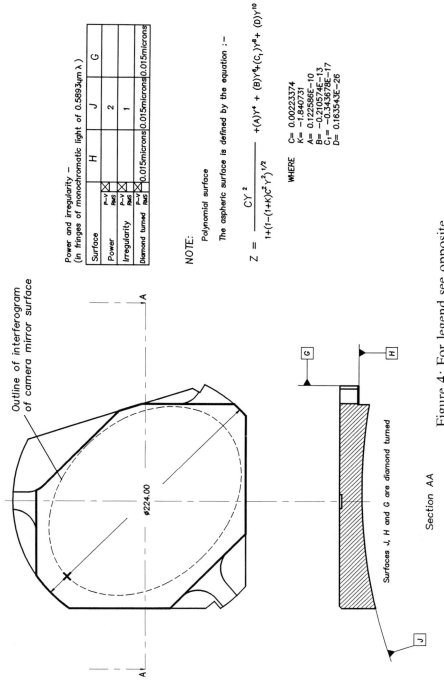

Power and irregularity –
(in fringes of monochromatic light of 0.5893μm λ)

Surface		H	J	G
Power	P-V	⊠	2	
	RMS	⊠		
Irregularity	P-V	⊠	1	
	RMS	⊠		
Diamond turned	P-V			
	RMS	0.015microns	0.015microns	0.015microns

NOTE:

Polynomial surface

The aspheric surface is defined by the equation :–

$$Z = \frac{CY^2}{1+(1-(1+K)C^2Y^2)^{1/2}} + (A)Y^4 + (B)Y^6 + (C_1)Y^8 + (D)Y^{10}$$

WHERE

$C = 0.00223374$
$K = -1.840731$
$A = 0.122586E{-10}$
$B = -0.210574E{-13}$
$C_1 = -0.343678E{-17}$
$D = 0.163543E{-26}$

Outline of interferogram of camera mirror surface

∅224.00

Surfaces J, H and G are diamond turned

Section AA

Figure 4: For legend see opposite.

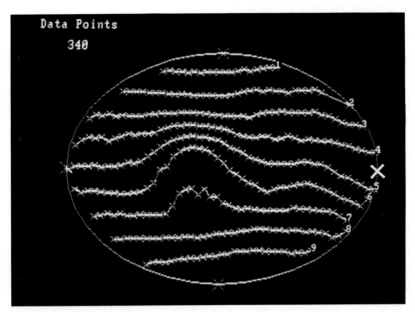

Figure 4. Design and testing of the camera mirror for CGS4. The interferogram shows the difference between the true surface and the diamond-cut surface.

central obscuration of the telescope and camera are taken into account (shown by the white circle) the surface precision over the illuminated region is better than 1 fringe at 0.659 μm. Subsequent analysis of this interferogram with Code V revealed a negligible degradation in image quality of the final camera at infrared wavelengths.

5.7. Mounting the optics

In order to maintain the tight tolerances on mounting the optical components of 25 μm decentre and 0.3 millirad tilt, high-precision manufacturing methods were used.

A casting was used as the support structure; these castings were machined referenced surfaces as well as the fixing holes and mounting shapes. As part of the optical-surface manufacture, reference surfaces were machined onto the single-piece blank which forms the mirror and what would in a conventional design be its mount. In this way mechanical machining and inspection are substituted for post-manufacture adjustment.

A second larger casting was used to take the bearings in which the grating and image rotator are carried. The optics-module casting was mounted to the main casting, again using precision-machined surfaces.

5.8. Structural considerations

The use of aluminium castings provides a homogenous stable structure. Use of all aluminium optics and structure gives an optical configuration which is homologous under thermal contraction. Elimination of the need for adjustment of optical components means that the instrument cannot become misaligned.

Low-loss gold coatings permit the use of two extra flat mirrors with little penalty. These make the instrument compact with two advantages:

low flexure with change of attitude,

lower surface area and therefore refrigeration requirement and weight.

5.9. Performance

In all cases the diamond-turned components and machined castings met the required tolerances or could be accommodated. The final alignment was determined both by optical means and by post-manufacturing inspection of the castings and component reference surfaces. In most cases the measurement techniques (using a 3-axis measuring machine) were found to have greater precision than the optical methods!

The resulting image quality of the spectrometer was tested at room temperature in laser light (0.589 μm) using an f/36 UKIRT telescope simulator. Over an ± 8 mm field in the detector focal plane the visible point-spread-functions were all inside 60 to 70 μm as predicted by Code V at these wavelengths and temperatures. Subsequently, on-telescope performance has verified the 30 μm 80% EED predicted for infrared wavelengths.

5.10. Efficiency

An estimate of the transmitted energy is shown in Table 1. Major losses are by the grating, the filter and the use of pierced optics to avoid off-axis configurations. An efficiency of 26% is predicted. An actual efficiency of 25% was measured on the telescope.

5.11. Future Trends

Diamond-turned optics have introduced to infrared the flexibility of using generalised aspheric surfaces; can this be extended to shorter wavelengths? Already qualities twice those specified and obtained for this instrument are now quoted for. Limitations are better understood and have to do with tool chatter and wear for example. A further factor of two will bring the technique into the grasp of visible optics designers.

Refinements of the design reported here, inspired by the need to instrument 8 m telescopes, permit a 120 arcsec long slit $\times$ 15 mm spectral field [Mountain *et al.*, 1991].

Table 1. CGS-8 through-put estimated by component at 2 μm

COMPONENT	COMPONENT MATERIAL/COATING	*	%
Cryostat Window	CaF$_2$ - OCLI coating		96
f/15 collimator	Gold coated A1 x 2	*	96
K Filter	OCLI broad-band measured at 77K		80
f/10 f-converter	Gold coated A1 x 2	*	96
Collimator Secondary	Gold coated A1	*	98
Collimator Primary	Gold coated A1	*	98
Folding flat 1	Gold coated A1	*	98
Grating	Gold-coated replica grating, Milton Roy 75 1/mm, 2.5 μm blaze at 2 μm in first order		70
Folding flat 2	Gold coated A1	*	98
Camera Mirror	Gold coated A1	*	98
Camera lens	ZnSe coated OCLI 1.0 - 5.5 μm		92
Hole in folding flat 2	Vignetting of system dominated by this stop		70
TOTAL THROUGH-PUT OF SYSTEM			29

*The minimum reflectivity of diamond-turned gold-coated aluminium at 1.0 μm and 3–5 μm measured by Diamond Electro-Optics Division of Contraves. Scattering is assumed to be neglible.

This is achieved by increasing the diameter of the hole piercing the flat feeding the camera mirror, which sacrifices some energy.

The energy budget is covered by using an anti-reflection-coated input window and by substituting an all-reflecting system for the construction of input collimator and f-converter.

6. Summary

Over a period of 15 years infrared optical design has developed as follows:

Initial designs for a 1.5 m telescope with single pixel detectors, using Mickey Mouse methods.

Progress to designs for 4 m class telescopes using anti-reflection-coated ZnSe lenses with the aid of simple ray-tracing.

Accommodation of fast optics and 2-D arrays with small pixels for full scale "visible-like" astronomical instruments by using diamond machining methods, aluminium mirrors and structures. Advanced ray-tracing and simulation software is essential at all stages of design through manufacture and test.

This has led to:

Success with CGS4.

We predict it will be possible to:

Extend the methods to the visible and to 8 m telescope infrared instruments.

References

[Erikson et al., 1984] Erikson E.F., Matthews S., Augason G.C., Houck J.R., Harwit M.O., Rank D.M., Hass M.R., *All-aluminium optical system for a large cryogenically cooled far infrared echelle spectrometer. Proc. SPIE, Vol. 509 (1984), p. 129.*

[Mountain et al., 1990] Mountain C.M., Robertson D.J., Lee T.J., Wade R., *An Advanced Cooled Grating Spectrometer for UKIRT. Proc. SPIE Conf., Tucson, Vol. 1235-04, Part 1 (1990), pp. 25-33.*

[Mountain et al., 1991] Mountain C.M., Robertson D., Atad E., Montgomery D., Pentland G., Pettie D., *A Cooled Grating Spectrometer: CGS 8. in United Kingdom Large Telescope Project Technical Report No. 20, Ed. Davies R.L., (1991).*

[Smith, 1981] Smith D.C., *Testing diamond turned aspheric optics using computer generated holographic (CGH) techniques. Proc. SPIE, Vol. 306 (1981), p. 112.*

[Wolfe & Zissis, 1985] Wolfe W.L., Zissis G.J., (Eds), *The Infrared Handbook. Environmental Research Institute of Michigan, Revised Edition (1985), Chapter 9.*

4
Optical Features of IRIS

Peter Gillingham *

Abstract

The AAO's infrared array camera/spectrometer, IRIS, was intended to max-
imise the usefulness of a near-infrared array with a minimum of delay, once the
array was delivered. It provides a large range of scales for direct imaging and
a few low-resolution spectroscopic modes with broad wavelength coverage; the
optical design lends itself to simple interchange between configurations. The
use of a Burch–Bowen two-mirror system for the widest direct field and the
adoption of special cross-dispersed transmission echelles were key elements.

1. Introduction

In planning the AAO's first instrument to use a two-dimensional infrared array,
versatility was a prime requirement. This followed both from the high cost of
arrays (with the consequence that we were likely to have only one array for some
years) and the desirability of coping with many different applications in any one
run on the telescope. Our first priority was to provide direct imaging with a good
range of scales, from one giving as wide a field coverage as possible with detector-
limited resolution to one which would seldom degrade the seeing. The next priority
was efficient low-resolution spectroscopy and the last was imaging with sufficient
resolution to take full advantage of the very best seeing.

2. The detector

Compared with telescopes at higher altitudes and lower temperatures, the AAT is less
competitive at longer wavelengths, so it was decided to buy a mercury–cadmium–
telluride array with cut-off at 2.5 μm. A Rockwell International (NICMOS2) array
with 128×128 0.06-mm square pixels was chosen and it was envisaged that there
might be a later upgrade to the NICMOS3 array with 256×256 0.04-mm pixels.

*Anglo-Australian Observatory, P.O. Box 296, Epping, NSW, Australia 2121.

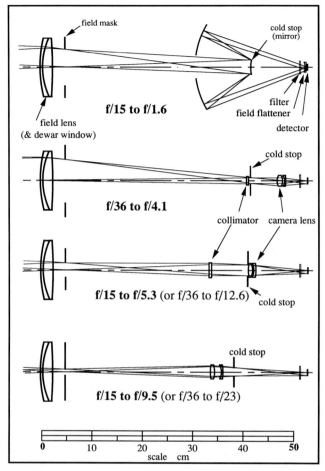

Figure 1. Configurations for direct imaging.

3. General optical layout

Fig. 1 indicates the configurations employed for direct imaging. Central to the design is a modification of the concentric mirror system studied in detail by Burch [Burch, 1947] and used in a microscope objective, and by Bowen [Bowen, 1967] in an astronomical spectrograph camera. This arrangement of two spherical mirrors plus a field-flattening lens gives good performance down to very fast focal ratios (e.g. f/1.6, as used here). The relatively large concave secondary mirror does not present a problem for interchanging optical configurations, as it can be fixed, while only the small primary mirror and field flattener are moved aside to introduce narrower field arrangements.

An important virtue of the "inverted mirror" focal reducer for this application is its compatibility with the alternative lens systems in respect of pupil position. The

40

convex primary mirror is placed at the pupil, where it has been re-imaged by the field lens. A simple lens, positioned at this pupil in place of the mirror and imaging the sky onto the detector, has a focal ratio of about f/5, which is appropriate for the medium field configuration. (So as to provide for spectroscopy with this medium field, the "camera" lens is preceded by a collimating lens and a space in which grisms can be interposed.) For the narrow field, corresponding to a final focal ratio of about f/9, the fact that the lens lies well ahead of the pupil is not a serious problem as the aberrations are still acceptable with a doublet.

It is necessary, towards the longer wavelength end of the spectral range, to exclude thermal radiation emanating from outside the telescope pupil (since the ambient-temperature black-body spectral radiance far exceeds that of the sky). So the Burch–Bowen primary is carefully sized to be slightly smaller in diameter than the pupil (i.e. it is made to act as a "cold stop"). Furthermore, the mirror is given a central obstruction slightly larger than the image of the telescope's central obstruction.

It was decided to exploit both f/15 and f/36 telescope foci. The f/15 focus is most compatible with the two-mirror focal reducer, and this telescope configuration has the advantage of being readily exchanged with the f/36 coudé (the two secondary mirrors are carried in the same top-end in a flip-over arrangement) so that sharing a night between a coudé and a Cassegrain f/15 instrument is quite practical. The slowly-converging beam at the f/36 Cassegrain focus is compatible with mounting a Fabry–Perot interferometer ahead of the field lens for higher resolution spectroscopy. It is feasible to use the lenses for the medium and the narrow field with both telescope configurations, so extending the range of imaging scales available.

Zero-deviation grisms are used for low resolution spectroscopy. The best combination of resolution and wavelength coverage is provided with the collimator and camera lens which are used only with the f/36 telescope top-end and produce a final focal ratio of f/4.1. Two cross-dispersed transmission echelles are installed, one covering the I and J atmospheric windows, the other H and K. In both cases, the resolving power $[\lambda/\Delta\lambda]$ corresponding to 2 pixels is about 400 and the maximum slit-length for no order overlap is about 15 arcsec. For long-slit work, the best arrangement is with the f/15 to f/5.3 lens system, for which single-order grisms, initially being provided for each of the H and K windows, will give R about 300 and allow a slit 76 arcsec long.

4. Details of individual optical features

4.1. Field lens

This lens acts as the vacuum-sealing window for the dewar as well as performing its optical role. It is an achromatic doublet of fused silica and calcium fluoride (fluorite), the elements being contacted with immersion oil. The negative silica element precedes the fluorite, for the sake of minimizing atmospheric exposure of the fluorite. Achromatic performance is required for the sake of throughput,

41

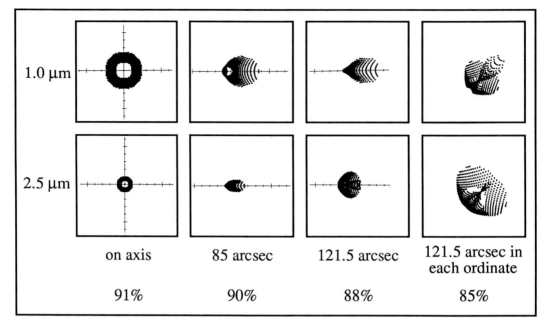

	on axis	85 arcsec	121.5 arcsec	121.5 arcsec in each ordinate
	91%	90%	88%	85%

Figure 2. Spot diagrams for the f/15 to f/1.6 configuration. 1000 rays were traced from the telescope pupil; the percentages indicate the unvignetted proportion of these rays reaching the array. The same compromise focus was used for the two wavelengths, at the extreme of the range used. The squares are 2 pixels (0.12 mm) per side.

rather than image quality at the detector. Near the long wavelength limit of about 2.5 μm, the lens must image the telescope pupil onto the cold stop, as mentioned earlier. A single-element field lens would have given serious vignetting at the shortest wavelength in the wide-field configurations.

A difficulty which was not fully appreciated until IRIS was in use was that any dust particle on the field lens, being at ambient temperature and close to the telescope focus, could, in the K window, appear as a relatively bright spurious star, even when its effective area corresponds to a very small fraction of a pixel's area. This problem has been minimized by introducing a forced flow of dry, filtered air around an annulus surrounding the lens, to maintain an upward flow throughout the time IRIS is on the telescope.

4.2.   Burch–Bowen system (f/15 to f/1.6)

The image quality calculated for this system is shown by the set of spot diagrams in Fig. 2. Schott SF57 glass was used for the field flattener as it combines a moderately high refractive index with low dispersion through the near infrared and, in the required thickness of just a few mm, it has good transmission. The curvature is under-corrected as full correction would have worsened the off-axis aberrations. Optimally-shaped baffles (squares with rounded corners) in the bore of the secondary mirror and just behind the primary mirror prevent any extraneous radiation from reaching the detector directly. On axis, about 9% of the radiation reaching the telescope focus is vignetted while about 15% is vignetted in the corners of the field. The AAT f/15 Cassegrain itself has a central obstruction of about 15% by area; the vignetting in a focal reducer of this nature would be a little more serious for a telescope with a smaller obstruction.

4.3. f/36 to f/4.1 lens system

To provide adequate resolution with as high a dispersion grating as the 63.4-deg echelle necessitated that the grating be mounted in a collimated beam. So this system comprises a collimating lens (a single element of silica) set just ahead of the pupil and a fluorite–silica doublet camera lens. With the collimated section long enough to accommodate an echelle, there is considerable expansion of the envelope of rays from the collimator to the camera lens in the direct imaging case. To maintain satisfactory imaging with a two-element lens over the full field, the camera lens aperture was restricted, giving about 20% vignetting at the corners of the array. In the spectroscopic mode, with much of the dispersion occurring at the second (echelle) surface of the grism, there is less expansion of the beam and no vignetting. Fig. 3 shows the direct imaging performance and Fig. 4 the performance in combination with the HK echelle.

The optimizing of this and the other lens systems was done (rather laborious-ly) using a specially-customised but relatively simple computer program. General purpose ray-tracing routines were incorporated into a program which would auto-matically optimise the bending of the second element of the camera doublet, while maintaining a designated focal ratio and other parameters as required. The designs were readjusted a little to take maximum advantage of curvatures for which tools were already available to the optical manufacturer. Of the several combinations of materials investigated, fluorite and silica gave the best overall control of aberrations, although some other combinations gave far smaller residual chromatic aberration.

4.4. Cross-dispersed echelle grisms

It was found that, by combining a transmission echelle at 63.4 deg with a first-order cross-dispersing grism, very useful combinations of wavelength coverage and resolving power could be provided. For the sake of minimizing reflection losses and simplifying the mounting of the grisms, the first-order ruling and the echelle were replicated onto opposite faces of the one prism.

In the case of the grism covering the I and J windows, BK7 glass could be used as the substrate. This has the important advantage of having a very good

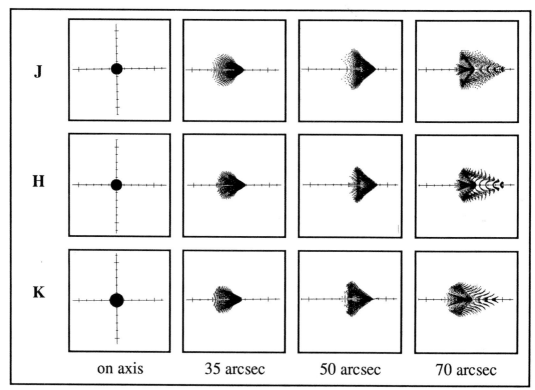

Figure 3. Spot diagrams for the f/36 to f/4.1 configuration used direct. About 1000 rays were traced for each of five wavelengths, representing the full passband for each atmospheric window. Slight telescope re-focus was made between windows. The squares are 2 pixels (0.12 mm) per side. There is no vignetting up to 50 arcsec and about 21% vignetting at 70 arcsec.

match in refractive index to the lower-index (so-called UV) resin used for replication by Milton Roy. Then rulings with groove angles matching the prism angles give maximum blaze efficiency along the optical axis and minimal reflection loss at the glass/resin interfaces. In the IJ case, existing master rulings (52.65 g/mm at 63.4 deg for the echelle and 210 g/mm at 26.75 deg for the cross-disperser, on prism faces inclined at 63.4 deg and 29.5 deg, respectively) provided a near-ideal echellegram format (see Fig. 5). Note that, to provide the best wavelength coverage and to align the most important orders approximately along the array rows, the grism dispersion is rotated considerably with respect to the array (but the slit is maintained parallel to the array columns).

For the HK grism, BK7 would not suit because of its absorption, and a search for a material with good transmission to 2.5 μm and a really close match to any of Milton Roy's resins was fruitless. The best combination found was fused silica

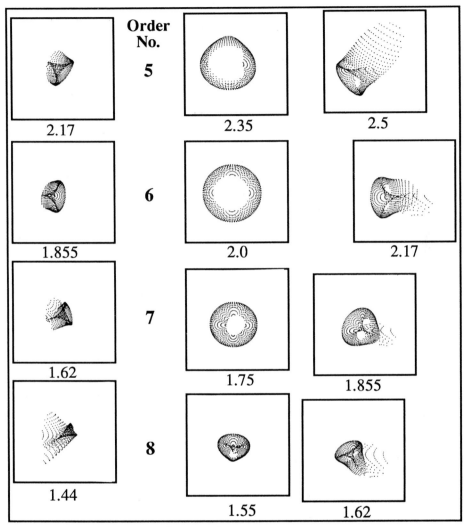

Figure 4. Spot diagrams for f/36 to f/4.1 lens with H-K echelle. Each square is 2 pixels (0.12 mm) per side. 1000 points are plotted for each spot. The positions of the squares roughly indicate the locations of the images within the array. Wavelengths (μm) are indicated under each spot. Focus was set to the best compromise for using this echelle and the I-J echelle.

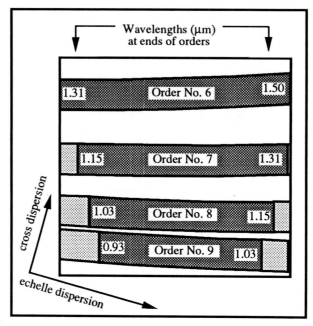

Figure 5. Layout of orders with 14 arcsec slit and I-J echelle.

(Heraeus Infrasil 301) with the UV resin. Fortunately, the penalty in reflection loss with a substrate of lower index than the resin is not severe, even for the interface inclined at 63.4 deg to the optical axis. For silica with the UV resin, there is about 0.6% reflection loss.

To set the maximum blaze along the optical axis, it was essential to allow for the deviations at the silica/resin interfaces and specify grating groove angles which differed very significantly from the prism angles (see Fig. 6). To avoid too serious an inefficiency from an effect akin to groove shadowing, it was also necessary, in the case of the echelle, to specify a roof angle on the grooves (corresponding to the angle of the ruling diamond) quite different from the standard 90 deg. This non-standard angle significantly increased the cost of the ruling and led to some delay before Milton Roy obtained a satisfactory diamond.

The echelle spectra suffer from a higher amount of stray radiation than is desirable, some of it concentrated in 'ghost' orders. It is likely that much of this arises from internal reflection and scattering, from the faces of the prism parallel to the optical axis (which were finished by grinding), of radiation sent by the cross-dispersing grism into unused orders. It is intended to add a baffle parallel to and immediately following the echelle surface to minimise the acceptance of this extraneous radiation. An additional measure being considered is to grind transverse grooves with saw-tooth cross-section into the offending prism faces.

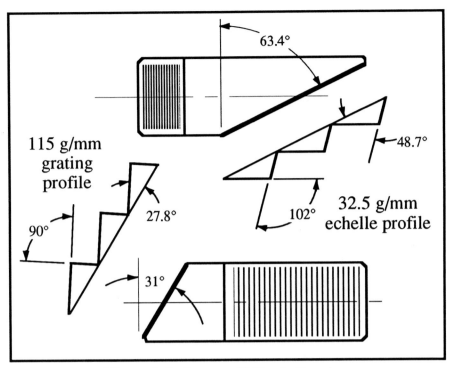

Figure 6. Rulings on H-K echelle grism.

5. Conclusion

Table 1 gives details of the various configurations already installed or planned from the outset. It demonstrates the wide variety of options which will be available when IRIS is complete. There are also plans to add polarimetric capabilities and some thoughts of using an external Fabry–Perot interferometer.

Through accepting optical aberrations which are, by some standards, gross, but which do not significantly degrade the resolution with the rather coarse infrared array, it has been possible to provide a very versatile imaging and spectroscopic instrument without undue optical complexity, cost, or delay in construction.

Acknowledgements

The ray-tracing routines employed were those written by Roderick Willstrop and often found to be very useful at the AAO, over the past several years. David Allen, as IRIS project scientist, tolerated the author's dabbling in lens design, even during the desperate month or two when it seemed we might never settle on a satisfactory simple lens for the echelle spectroscopy. Many other AAO staff helped build IRIS;

Table 1. IRIS configurations, available and planned.

Telescope focal ratio	Direct imaging			Spectroscopy		
	Final focal ratio	Pixel size (arcsec)	Field (arcsec)	Range	$R= \lambda/\Delta\lambda$	Slit length (arcsec)
f/15	1.6	1.9	240 × 240			
	5.3*	0.59*	76 × 76*	H, K*	~300*	76*
	9.5[†]	0.33[†]	42 × 42[†]			
f/36	4.1	0.78	100 × 100	I+J, H+K	~400	15
	12.6*	0.25*	32 × 32*			
	23[†]	0.13[†]	16 × 16[†]			

*Medium field lens, used with both telescope top ends, to be available from early 1992.
[†]Narrow field lens, used with both telescope top ends, deferred to mid-1992 or later.

in particular, Allan Lankshear, Paul Lindner and Gordon Shafer, working at Siding Spring, designed, machined, and assembled the very many components needed to hold, cool, and move the optics.

References

[Bowen, 1967] Bowen I.S., *Q. J. Roy. astr. Soc., Vol. 8 (1967), p. 9.*

[Burch, 1947] Burch C.R., *Proc. Phys. Soc. London, Vol. 59 (1947), p. 41.*

5

The FLAIR II system
at the UK Schmidt Telescope

F. G. Watson,[*]*P. M. Gray,*[†]*A. P. Oates*[‡]*and T. R. Bedding*[§]

Abstract

Since 1988, the Anglo-Australian Observatory's 1.2-m UK Schmidt Telescope has operated a successful multi-object spectroscopy service using a version of the FLAIR optical-fibre feed system known as PANACHE. Developed principally for galaxy redshift measurements, the system has routinely provided velocities for objects with $B \sim 17$ mag. The full potential of the telescope for carrying out multi-object spectroscopy has not been realised, however, because of PANACHE's limited number of fibres and lengthy fibre-reconfiguration time. To overcome these limitations, a second-generation instrument has been built with ~ 100 fibres, rapid turn-round of target fields, and a new, optimised spectrograph. This FLAIR II project has been undertaken with SERC funding and AAO manpower, and is now substantially complete.

1. Introduction

It is almost a decade since the potential of fibre-coupled multi-object CCD spectroscopy over the 40 square-degree field of the 1.2-metre UK Schmidt Telescope (UKST) was first investigated [Dawe & Watson, 1982; Dawe & Watson, 1984]. The development of the FLAIR (Fibre-Linked Array-Image Reformatter) system from the original concept to the first operational prototype (1985) and thence to the successful PANACHE (PANoramic Area-Coverage with High Efficiency) version in 1988 was a lengthy process that has already been well-documented in the literature [Watson, 1986; Watson, 1988; Watson *et al.*, 1990].

FLAIR was the first operational multi-fibre system on a Schmidt-type telescope, and the first multi-fibre system on *any* telescope to feed a floor-mounted spectrograph. Surprisingly, it remains the only working instrument of its kind on a Schmidt,

---

[*]Anglo-Australian Observatory, Coonabarabran, NSW, Australia 2357 *and* Royal Observatory, Blackford Hill, Edinburgh EH9 3HJ, UK.

[†]Anglo-Australian Observatory, PO Box 296, Epping, NSW, Australia 2121.

[‡]Royal Greenwich Observatory, Madingley Road, Cambridge CB3 0EZ, UK.

[§]School of Physics, University of Sydney, Sydney, NSW, Australia 2006.

49

although plans do exist to install fibres on both the 1.3-m Tautenburg Schmidt and the proposed Chinese 1.5-m Schmidt. At the UKST, FLAIR with PANACHE has, since 1988, operated as a service instrument at the level of approximately three observing nights per lunation, alongside the telescope's photographic work. Its main role has been in redshift determination for galaxies with $B \leq 17$ mag (e.g. [Parker & Watson, 1990; Watson et al., 1991]) and, by 1991, spectra of some 2400 galaxies (along with about 400 stars and 100 quasars) had been amassed for a large number of users.

The most striking aspect of these figures is that they could have been very much higher had the system been capable of accepting more fibres and/or had a more rapid field turn-round time. There are, for example, approximately 200 galaxies per Schmidt field down to $B = 17$, of which only 30 could be sampled simultaneously by PANACHE (allowing for 6 sky fibres). With only one fibre plateholder available and a field-change time approaching a full working day, the rate of progress in obtaining redshifts was clearly very limited. For this reason, the early redshift surveys were carried out at a sampling ratio of 1-in-6 [Watson et al., 1991; Watson et al., 1992a].

In order to speed the rate of progress of a new 1-in-3 redshift survey [Broadbent et al., 1992; Watson et al., 1992a] a makeshift solution was adopted which involved running two separate 36-fibre feeds from the single PANACHE plateholder to two separate spectrographs (each with its own CCD camera) to provide a 72-fibre capability [Watson et al., 1990; Oates, 1990]. This failed to address either the problem of positioning this number of fibres in the plateholder or the defects of the rudimentary prototype spectrograph. Clearly, the main requirement was for a second-generation instrument that would eliminate the operational deficiencies of PANACHE, and provide the UKST with a true common-user multi-object spectroscopy system for the 1990s.

Work on this system, FLAIR II, commenced in 1989 and, with the exception of the supplementary plateholders, is now (Jan 1992) complete. Recent papers by Watson et al. [Watson et al., 1992b] and Bedding, Gray and Watson [Bedding et al., 1992] describe FLAIR II in some detail; here we compare its design and performance with those of its predecessor.

2. FLAIR with PANACHE – something of a misnomer?

Table 1 gives a summary of the more important characteristics of the PANACHE system. The 36 fibres were terminated at their input ends with 2-mm square 90-deg micro-prisms, and were positioned in the focal surface by an optical alignment method. This was a simple and inexpensive way of achieving the ~ 10 μm accuracy required by virtue of the fine plate-scale (67 arcsec/mm) of the telescope, and the basic principle has been retained for FLAIR II. In the set-up procedure, the virtual image (in its micro-prism) of each back-illuminated fibre is aligned with a selected image on a copy-plate of the target field, and the fibre-ferrule tacked in place with a non-permanent UV-curing polymer. When all the fibres are positioned, the plate is

Table 1. FLAIR with PANACHE

No. of fibres	36
Fibre diameter	100 μm≡6.7 arcsec
Fibre length	10.7 m
Fibre type	high-OH$^-$ (blue-trans.)
Total reconfiguration time (incl. fid.)	~4 hours
Load time into telescope	30 minutes
No. of fields per night	1
Spectrograph optics	Pentax SLR lenses
Slit length	5 mm
Beam diameter	25 mm
Effective CCD size	133×289 binned pixels
On-line storage capacity	25 binned frames

tensioned in its plateholder to conform to the telescope's focal curvature, and loaded into the telescope in a similar way to the photographic plateholders.

A satisfactory acquisition system was developed early in the history of FLAIR [Watson, 1986]; its basic features have been retained throughout. A 1.1-mm image guide (coherent fibre-bundle) fitted with an input-end micro-prism brings to an intensified TV camera a 1.2-arcmin portion of the telescope field, in the vicinity of an acquisition star. Once this star has been acquired and centred, the plateholder is driven in rotation until five fiducial-star fibres, identical to the main fibres except in diameter (33 μm ≡ 2.2 arcsec), also light up in the TV image indicating that the field is properly aligned. The telescope's autoguiding system can then be engaged.

Although the PANACHE fibre-positioning and field acquisition techniques worked well in terms of the results they produced, the rudimentary design of the mechanical components made for extremely cumbersome procedures that were not suitable for expansion to a greater number of fibres. The fibre-positioning (x, y) table was equipped with only a visual microscope and manual controls, and placed considerable strain on the operator. Fibres had to be positioned with a field-dependent offset, which arose because the fibre ends were not optically conjugate with the plate surface [Watson *et al.*, 1990]. Excess lengths of fibre trailing over the front surface of the plate had to be laboriously coiled and taped down before loading into the telescope, while the loading process itself required several excursions to the telescope's interior by the operator, a generally dangerous procedure.

At their output ends, the PANACHE fibres formed a slit in the focus of a rudimentary spectrograph made from two Pentax SLR camera lenses with interchangeable 50-mm square plane reflectance gratings. The lenses performed well except in

Table 2. FLAIR II – initial configuration

No. of fibres	92
Fibre diameter	100 μm≡6.7 arcsec
Fibre length	11.5 m
Fibre type	high-OH$^-$ (Blue-trans.)
Total reconfiguration time (incl. fid.)	∼3 hours
Load time into telescope	5 minutes
No. of fields per night	2
Spectrograph optics	all-Schmidt
Slit length	20 mm
Beam diameter	150 mm
CCD size	400×578 pixels
On-line storage capacity	∼ 240 frames

the blue ($\lambda < 4400$ Å), and the system provided a range of reciprocal dispersions from 373 to 54 Å mm^{-1}. The spectrograph was mounted on a vibrationally-isolated table in the dome; the benefits of this arrangement were reflected in the extreme stability of the data. The spectrograph camera was built around an EEV P8603B CCD image-sensor, dye-coated for enhanced blue sensitivity, and operated in slow-scan mode at a temperature of 150 K. An ancient DEC PDP 11/23 minicomputer with very limited hard-disk storage controlled the camera through CAMAC.

The main defects of the spectrograph were its poor blue performance and a geometry that resulted in a gross mismatch between the fibre diameter and the CCD pixel size (so that the detector had to be operated with on-chip binning in both dispersion and slit directions, thereby wasting area and resolution). These both had a deleterious effect on the instrument's sensitivity for galaxy redshift observations, first because the H and K lines of Ca II were difficult to detect, and second because red data at the required resolution (to pick up any H$_\alpha$ emission lines) needed separate exposures. Nonetheless, the system yielded velocities with an 80% success rate for galaxies of all types with $B \leq 16.8$ mag (with an accuracy of ∼150 km/s) in ∼ 12000 sec, and with a 100% success rate for galaxies exhibiting H$_\alpha$ emission with $B \leq 17.5$ mag (to ∼60 km/s).

3. FLAIR II – living up to the name …

The design philosophy of FLAIR II has been to retain the well-tried basic principles of FLAIR/PANACHE, but to engineer the system so as to permit a much more streamlined operation with a greater number of fibres. Thus, each of the individual components of FLAIR II is a major development of its counterpart in the earlier system. Details of the new instrument are summarised in Table 2.

52

3.1. Plateholders and fibre feeds

The FLAIR II plateholders incorporate fibres that retract into the body of the plateholder, eliminating the difficulty with excess lengths of fibre, and reducing the amount of handling to which the fibres are subjected. Significant problems were encountered in the design because of the very limited space available (see [Watson et al., 1992b]); the finished plateholders include not only the retraction system, but also integrated rotation-drive and readout units. This simplifies the telescope loading procedure, dramatically reducing the amount of time required. The plateholder capacity is 152 fibres; initially there will be two, each equipped with $92 \times 100 \mu m$ fibres for galaxy redshift surveys at two fields per night. The first of these is now completed. At a later date, it is possible that two further plateholders with 50 μm (3.3 arcsec) feeds will be commissioned for quasars and stars.

Improvements to the fibre feeds include new input-end ferrules fitted with extended micro-prisms to render the fibre-end conjugate with the plate surface, thus eliminating the field-curvature offset. (Likewise the image guide has a new micro-prism design; in order to provide additional optical thickness together with an image-reversal it has the unusual form of a Porro-prism of the second class with an exit-face extension.) All the fibre ferrules are now fitted with Teflon pads to aid clean release from the copy-plate.

3.2. "AutoFred" fibre positioner

In order to speed positioning of the larger number of fibres, the positioning table has been equipped with a semi-automatic fibre positioner (known as "AutoFred"), consisting of a robot under the control of a PC, together with a CCD camera and framegrabber [Bedding et al., 1992].

In use, the operator moves the plateholder to bring the target object into the field of view of the camera, and identifies the target with a cursor. With the fibre-ferrule loaded into a pneumatically-operated gripper, the computer locates the backlit fibre on the TV display, and moves it to align the fibre image with the target object to within a predetermined error (usually 1 pixel, or ~ 8 μm). Once the fibre is correctly positioned, the system opens the UV shutter to fix the fibre in place, releases the gripper, and writes the fibre and object identifications to a log file before proceeding to the next fibre.

"AutoFred" has significantly increased the fibre-positioning rate and has brought major improvements in accuracy and consistency, rendering the job of positioning ~ 100 fibres a relatively straightforward task.

3.3. FISCH spectrograph and CCD system

In order to satisfy the spectrograph requirements of a fast, wide-field collimator and a large collimator/camera focal-length ratio, the design of the optics mimics that of the telescope itself. Both collimator and camera are Schmidt systems, the former working at f/2.1 (the approximate focal-ratio of the telescope beam after degradation by the fibres), and the latter at f/1.3 [Watson *et al.*, 1990; Watson *et al.*, 1992b]. The spectrograph is known as FISCH (for FIbre-SCHmidt), and its optics are adapted from two redundant spectrograph cameras on long-term loan from the Royal Greenwich Observatory.

The collimator is fed by a fibre-slit at its internal focus. This is polished flat (rather than convex) and has the fibres mounted parallel (rather than fanned to the focal-curvature) with no serious loss of image-quality or increase in vignetting. The camera Schmidt-corrector forms the front window of a CCD cryostat containing the camera mirror together with the image-sensor itself (again a P8603B device). A field-flattening lens is fitted, and the detector is maintained at its operating temperature by means of a cold-finger from the integral liquid-nitrogen dewar and a thermostatically-controlled heater behind the chip.

FISCH has a camera/collimator separation of 45 deg, and utilises 200 mm × 150 mm plane reflectance gratings of the type used in the AAO's RGO spectrograph (which are thus available for use at the UKST in a wide range of blaze-angles and groove-spacings). A selection of the options available is shown in Table 3.

Table 3. Typical FLAIR II grating configurations

Grating	Blaze direction	Reciprocal dispersion (Å/mm)	Instrumental resolution (Å)	CCD binning	CCD resolution (Å/pixel)	Spectral range (Å)
250B	cam.	268	12.6	× 2	11.8	3400
250B	coll.	286	11.3	× 1	6.3	3620
600V	cam.	108	5.1	× 2	4.7	1360
600V	coll.	122	4.5	× 1	2.7	1540
1200B	cam.	49	2.6	× 2	2.2	620
1200B	coll.	62	2.1	× 1	1.4	790

Of all the components of FLAIR II, the FISCH spectrograph has contributed most to improved sensitivity, by virtue of the enhanced blue performance and better matching to the CCD. Since the new spectrograph was introduced, a gain in sensitivity of about 0.3 mag has been achieved in galaxy redshift observations.

The final major component of FLAIR II is a new CCD system. The old PDP 11/23 computer has been replaced with a 25 MHz 80386 PC running GEM windows, communicating with the CCD sequencer via a 48-bit I/O card and an opto-isolated CCD interface [Oates, 1990]. New software provides a much more efficient environment for image acquisition and processing [Oates, 1991], while a

172-Mbyte hard disk has dramatically increased on-line image storage capacity. The PC is connected by Ethernet to the AAO's Vaxcluster, enabling files to be transferred to the bigger machines for data reduction and output to tape.

4. Conclusion

FLAIR II has been introduced in phases, each new component replacing its counterpart in the older system to allow adequate commissioning time. Chronologically, the new spectrograph and CCD system were introduced first (Feb 1991) while, at the time of writing, the first of the plateholders (the final new component of the system) is undergoing commissioning.

Unlike its predecessor, which was operated in service-mode by UKST staff, FLAIR II will be run as a conventional common-user system, with astronomers expected to be present at the telescope. However, it is likely that responsibility for handling the hardware of the system, including fibring-up the plates, will remain with UKST support astronomers.

There appears to be no reason why FLAIR II should not now go on to fulfil the promise displayed by the PANACHE system. With the FLAIR time-allocation set to increase to five nights per lunation, the possibility of observing two fields per night and the greater number of fibres per field, there is a real possibility that the new system will increase the FLAIR data-collection rate by a factor of ten.

5. Acknowledgements

It is a pleasure to thank the staff of the UKST, the Anglo-Australian Telescope and the UKST Unit of the Royal Observatory, Edinburgh, for their support in the operation of FLAIR. Particular thanks go to Malcolm Hartley for his valuable assistance in recent months. The contribution of the following to the development of FLAIR II is acknowledged with gratitude: Ian Bates, Eric Coyte, Robert Dean, Michael Kanonczuk, Allan Lankshear, Paul Lindner, Don Mayfield, André Porteners, Doug Pos and Dennis Whittard. The project has been funded largely by SERC, and has always been blessed with the enthusiastic support of the Directors of AAO and ROE.

References

[Bedding *et al.*, 1992] Bedding T.R., Gray P.M., Watson F.G., in *Fibre Optics in Astronomy II, ed. Gray P.M., Astron. Soc. Pacif. Conference Series, 1992, in press.*

[Broadbent *et al.*, 1992] Broadbent A., Hale-Sutton D., Shanks T., Watson F.G., Oates A.P., Fong R., Collins C.A., MacGillivray H.T., Nichol R., Parker Q.A., in *Digital Optical Sky Surveys, ed. MacGillivray H.T., Kluwer, 1992, p. 389.*

[Dawe & Watson, 1982] Dawe J.A., Watson F.G., *Proposal for a radical extension in the use of the UK Schmidt Telescope. Internal Report, Royal Observatory Edinburgh, 1982.*

[Dawe & Watson, 1984] Dawe J.A., Watson F.G., in *Astronomy with Schmidt-type Telescopes, ed. Capaccioli M., D. Reidel, Dordrecht, 1984, p. 181.*

[Oates, 1990] Oates A.P., *Proc. SPIE, Vol. 1235, 1990, p. 272.*

[Oates, 1991] Oates A.P., *FLAIR CCD system user guide., Anglo-Australian Observatory Manual TM 8.1, 1991.*

[Parker & Watson, 1990] Parker Q.A., Watson F.G., *Astron. Astrophys. Supp. Ser., Vol. 84 (1990), p. 455.*

[Watson, 1986] Watson F.G., *Proc. SPIE, Vol. 627 (1986), p. 787.*

[Watson, 1988] Watson F.G., in *Fiber Optics in Astronomy, ed. Barden S.C., Astron. Soc. Pacif., Vol. 3, 1988, p. 125.*

[Watson et al., 1990] Watson F.G., Oates A.P., Gray P.M., *Proc. SPIE, Vol. 1235 (1990), p. 736.*

[Watson et al., 1991] Watson F.G., Oates A.P., Shanks T., Hale-Sutton D., *Mon. Not. R. astr. Soc., Vol. 253 (1991), p. 222.*

[Watson et al., 1992a] Watson F.G., Broadbent A., Hale-Sutton D., Shanks T., Parker Q.A., Oates A.P., *Proc. Astron. Soc. Aust., Vol. 10 (1992), p. 12.*

[Watson et al., 1992b] Watson F.G., Gray P.M., Oates A.P., Lankshear A., Dean R.G., in *Fibre Optics in Astronomy II, ed. Gray P.M., Astron. Soc. Pacif. Conference Series, 1992, in press.*

6
Focal Length Converters

Richard G. Bingham [*]

Abstract

Many of the auxiliary optical systems used with telescopes have the effect of changing the focal length. This paper firstly points out the several useful functions which such optical systems can perform. Numerous optical layouts are next outlined, with notes on their applications. Various practical examples, mostly not previously published, are described.

1. Introduction

In the design of a telescope with a Cassegrain or Nasmyth focus, the choice of f-number often involves the need for a reasonably small secondary mirror and central obstruction. Also, the convergence of the beam should not much degrade the performance of devices such as filters, beamsplitters, prisms, polarisation devices and collimators. The size of such components will also be affected by the choice of f-number. In addition, there may be a need for some field of view where aberrations with no additional corrector are small. These considerations usually imply a relative aperture of about f/8 to f/15. A longer path to the focus may be provided at f/30, although the field of view may be restricted and collimators need to be longer or more complicated. At the fast end of the range, f/6 has been suggested for input to optical fibres, but requires a relatively large secondary mirror or very fast primary mirror.

Although such considerations may define the telescope, different focal lengths overall are usually required for the functions of the optical instrument. The functions include those of the acquisition system and autoguider as well as the auxiliary instrument, and the final relative aperture may be around f/1 to f/4. Therefore we use additional optics which change the focal length, for various reasons listed below. Often the auxiliary instrument consists basically of optics which re-image the Cassegrain focus, albeit incorporating optical devices such as etalons or gratings. Therefore "focal length converters" or "focal reducers", etc., have a general relevance to optical instruments for astronomy. Special reference is due to earlier work at Haute Provence and elsewhere [Courtès, 1972].

---

[*]Royal Greenwich Observatory, Madingley Road, Cambridge CB3 0EZ, UK.

Thus there are various aims in the design of focal length converters, such as:

Changing the f-number to match the input of an existing instrument or design

Moving the focus ...

... or keeping the position of the focus the same despite the insertion of the converter

Increasing the field of view on a given detector, e.g. for acquisition

Matching the seeing to a detector with the required sampling

Increasing photographic speed

Compensating for aberrations arising in the telescope

Creating a space for further optical devices

Providing a collimated beam for optical devices

Correctly positioning a pupil to suit the auxiliary instrument (A "pupil" may be a real or virtual image of the aperture stop, which is usually the edge of the primary mirror or a chopping secondary mirror.)

Setting the size of a pupil

Maximising the throughput of a spectrograph in the presence of variable seeing

Achieving the usual optical specifications such as high throughput, wavelength coverage and small aberrations

Designing the lens to fit the box – referring to mechanical constraints.

2. Some basics

Fig. 1a shows the optics whose focal length is to be modified as a "black box" camera C, indicating that it can have various forms. Most of the converters can be applied to a reflecting telescope, which would be one such form. Figures 3b onwards discard the black box in order to concentrate on the area of the focus and auxiliary instrument.

Focal length as a concept is useful for imaging systems which can be focussed to infinity. Thus the focal length of the camera shown in Figure 1a is given by [Welford, 1986]:

$$f' = -h/u'$$ (1)

where f' is the focal length, h is the height from the axis of a ray in star space, and u' is the semi convergence angle of the beam. In this equation, u' may be considered

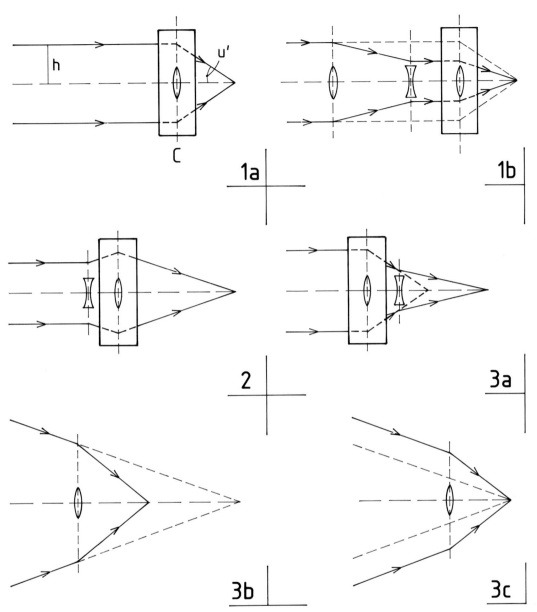

Figures: **1a.** A "black box" camera C with a focal length of $-h/u'$. **1b.** Type 1. An afocal lens combination in front of the black box camera, increasing its focal length. **2.** Type 2. A negative-power front lens, increasing the focal length. **3a.** Type 3. A negative rear lens, increasing the focal length. **3b.** A positive rear lens, reducing the focal length. **3c.** A positive rear lens, reducing the focal length, but with the image position kept constant, typically by refocussing a Cassegrain telescope.

as a small angle. (The primes indicate quantities applying on the image side rather than the object side. The minus sign conforms with sign conventions, in particular that angles are positive if measured anticlockwise.) If the convergence angle is not small, it is necessary to be cautious about aberrations. Equation 1 is applicable if we consider the paraxial region only. For finite rays in the full aperture, in the case of zero linear coma, u' in equation 1 can be replaced by $\sin U'$ where U' is the finite convergence angle. (This form could be used for computing U'. It does not apply exactly to the Cassegrain system, which has non-zero coma, but the differences are usually small.)

Thus changing the focal length of the combination of the instrument and a given telescope is a matter of changing the convergence angle of the focussed axial ray pencil. The *front* ray height h is usually considered constant when it applies to the aperture radius of the primary mirror.

There is an alternative approach if the value of h is changed, as it could be if the beam was recollimated (Figure 1b). (There is a potential application in auxiliary instruments.) The effect is constrained by Lagrange invariant H, which is preserved in an optical train [Welford, 1986];

$$H = -nh\beta = n'u'\eta' \qquad (2)$$

where n and n' are refractive indices, h is the beam radius in a collimated (star) space, and β is the semi field angle in that space, giving rise to an image height η'. The focal length may be increased by inserting lenses to reduce the incidence height h in a collimated space, as illustrated in Figure 1b. This may be understood either from equation 1 (from the reduction in the convergence angle u' of the subsequent ray pencil) or by the magnification of the field angle β, indicated by conserving H while reducing h in equation 2. The action is the reverse in the case of a beam expander, which reduces the focal length.

3. Types of converters

This section describes a number of optical layouts which represent many of the practical forms of focal length converter. They are simplified in this section. Thus Figures 1–7 do not include various features necessary for the correction of aberrations – lenses shown conventionally as small symbols on the optical axis will appear as compound lenses in the practical examples.

The examples will use spherical optical surfaces rather than aspherics, with one or two obvious exceptions which are pointed out.

Type 1 – Front beam expander or reducer. As sketched in Figure 1b, an afocal combination of optical elements, changing the aperture of a collimated beam, can be placed in front of a camera and increases the focal length if it reduces the beam aperture. The focus is not moved in relation to the black-box camera. The additional optical elements are a Galilean or other telescope.

Type 2 – Front power change. A simple but limited method of, say, increasing the focal length of a small camera is to place a negative lens close to a positive lens to reduce the total optical power. The negative-power front lens illustrated in Figure 2 is placed close to the normal camera lens, which then requires a large focus adjustment, for example with the insertion of a mechanical spacer. Front converters are, of course, impractical for large telescopes.

Type 3 – Converter lens in a converging beam. A positive or negative lens (Figures 3a and 3b) used on the rear (exit) side of the optics is relevant where the main camera is a large telescope. It tends to be a low power device, compared to a re-imaging lens following the focus, and may be a fairly simple lens group when the field is small. The focus is moved in the sketches, but the designer may restore the focus position by adjustment of the secondary mirror of a Cassegrain system. This gives the effect indicated in Figure 3c.

Type 4 – Parfocal converters in a converging beam. With the use of two separated components in a converter (Figure 4a), the magnified or diminished image may be placed in focus on the original plane, and the conversion is said to be parfocal. The assembled device is therefore insertable when required without moving the detector or refocussing the telescope. Figure 4b shows an alternative arrangement following the focus, which can be used inside a spectrograph [Diego & Walker, 1985; Bingham, 1979]. [Diego & Walker, 1985; Bingham, 1979]. Figure 4c shows a system giving the reverse effect of Figure 4a.

Sometimes such arrangements can usefully be made parfocal with respect to a second object at a different position. This second object may conveniently be a pupil. Thus the new image may be formed in the original plane while also retaining the position of the exit pupil. This is a useful feature when a retractable converter feeds another optical system, or if a telecentric character is to be retained with the addition of the device. This will be referred to as a "doubly parfocal" converter.

Type 5 – Parfocal relayed image. Figure 5 shows an alternative to Figure 4c. It relays the image. In Figure 5, the focus is pulled back by the first component to create space for a re-imaging lens in front of the original focal plane. With the positive front element as shown in Figure 5, a real pupil from a large telescope can be formed not too far from the second lens, reducing its necessary diameter. In Figure 4c, the combination of negative and positive lenses is attractive at first sight for correcting aberrations, and the second lens has less power. However, this is less appealing when the field size is significant, as the second lens is far from a pupil.

Type 6 – Post-focus relay. Figures 6a and 6b show a further approach in which the original real image is formed unchanged and subsequent optics are used to re-image it with a different plate scale. It will usually be an advantage to include a field lens, thus reducing the diameter of the re-imaging section, by placing a real pupil close to or within it. Figure 6a includes a field lens F and a re-imaging or relay lens R, and the indicated rays apply to an off-axis field point. It is commonly assumed that the field lens (as its name implies) will be close to a focal surface, as shown in Figure 6a, in order to reduce some of the aberrations it contributes. That is a useful guide, but in practice the requirement is relaxed as other optics are

61

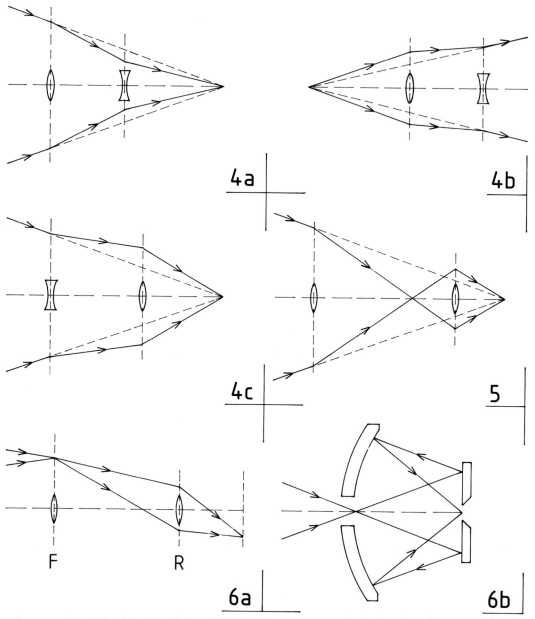

Figures: **4a.** Type 4. A parfocal device of two lens elements, increasing the focal length. **4b.** As Figure 4a, but within a re-imaging system. **4c.** As Figure 4a, now reducing the focal length. **5.** Type 5. This is often a better way to reduce the focal length with a parfocal device. **6a.** Type 6. A re-imaging system with field lens F and relay lens R. **6b.** A two-mirror re-imaging system.

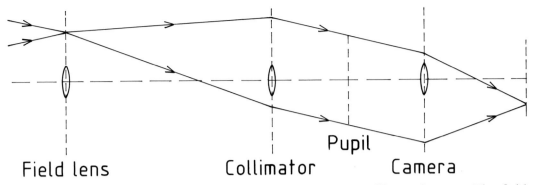

Field lens   Collimator   Pupil   Camera

Figure 7. Type 7. A re-imaging system incorporating a collimated space. The field lens is typically close to the Cassegrain focus. The collimated space typically includes an etalon, transmission grating, adaptive optics device or a polarisation analyser close to the pupil.

designed at the same time. Then aberrations introduced by the field lens may be compensated elsewhere, such as in the relay section of Figure 6a, which will normally be a compound lens. Figure 6b illustrates a reflecting variant of the post-focus relay. The mirror surfaces are aspherical. It typically also requires lens elements close to either image position [Bingham, 1978].

Type 7 – Collimated spaces, instruments. If the re-imaging part of the scheme in Figure 6a is split into collimator and camera sections, a layout such as in Figure 7 is obtained. A grating or Fabry–Perot etalon may be inserted in the collimated space. The field lens and collimator also have the function of placing a pupil image on the grating or etalon, exploiting the size of those devices. The field angle in the collimated space is readily calculated using equation 2.

4. Examples

The principles given above are illustrated here by various examples designed by the author or as otherwise acknowledged, and mainly produced by IC Optical Systems Ltd, Optical Surfaces Ltd or Grasbey Specac Ltd. Broad indications of aberrations are given in most cases – these are total geometrical image spreads. To avoid the necessity for several tables of data which would be of limited interest, the exact design details are not given. However, the lens drawings and text show all the main points and some glass types are stated, using the Schott nomenclature. Another lens designer can therefore reproduce the functions of these lenses if required. Also, it is hoped that designs will be available from the author for some time to come. The majority of applications described are in use on telescopes of the Isaac Newton Group, La Palma [Sánchez & Wall, 1991]: the 1-m Jacobus Kapteyn (JKT), the 2.5-m Isaac Newton (INT), and the 4.2-m William Herschel (WHT). Other applications are used on the 3.9-m Anglo-Australian (AAT) telescope and the 1.88-m telescope of the South African Astronomical Observatory.

These schemes arise from typical problems in astronomical and other optics work and may have been used in various places; this paper uses the results of such work at the Royal Greenwich Observatory to assemble some of the principles and examples for future reference. These lens types and examples have all arisen in the course of work for specific projects, and other methods will surely be developed in response to future demands.

It may be worth noticing that commercially available lenses, although sometimes appropriate, can raise difficulties in critical applications. Manufacturers release only rudimentary data, making it impossible to raytrace the lens or to optimise other optics in the system fully with regard to aberrations. A chain of two or more such lenses gives rise to problems with the accumulation of both aberrations and vignetting from one lens to the next, aggravated by problems of pupil imagery. Production of the lens may cease before the project is finished, and examples can differ. A lens mechanism which is difficult to service may fail mechanically. So when such a lens seems only barely adequate, it will be worth considering a special design. Its performance may relax tolerances in other areas, including focus, or may increase the wavelength coverage. Zoom lenses are little used in astronomy, probably because of the problems discussed here, their general complexity and possibly questions regarding their entrance pupils.

Examples of types 1 and 2. The use of converter lenses in front of the camera (Figures 1b and 2) is exemplified by some accessories for commercial lenses, principally for small photographic and video cameras. Some commercial lenses of type 2 appear to consist of a single weak lens element with an aspheric surface. No examples in astronomy are given here, though the principles are potentially useful. Type 1 (Figure 1b) could be used within the collimated space of a collimator–camera system, leaving the original lenses in place. It would result in a more complicated total system than using a complete interchangeable lens. However, an application of type 1 might be to modify the plate scale of an imaging system without interfering with the main optics. That might be attractive in a space project. The mechanical tolerances for inserting the converter as a whole in a collimated space are greatly relaxed, compared to those for a fast lens. The converter may be complicated and is best designed in the same exercise as the original lenses.

Examples of type 3. The field of view of the acquisition TV system for the f/15 focus of the 1-m Jacobus Kapteyn Telescope is increased by a positive lens in the converging beam, at 365 to 588 nm. Fig. 8 shows the lens which converts the f/15 focus to f/6.9. A simple, low cost solution was demanded, in a small space rather close to the TV sensor (fitting the box!). A cemented doublet provides the necessary power, and a meniscus serves to adjust the astigmatism. Astigmatism is balanced against Petzval field curvature to flatten the surface containing the disks of least confusion [Welford, 1986]. The flint meniscus element also controls the lateral colour: it may be noticed that the flint–crown–flint structure is quite different from the Cooke triplet. At the corners of the 25-mm diagonal TV field, the aberrations

64

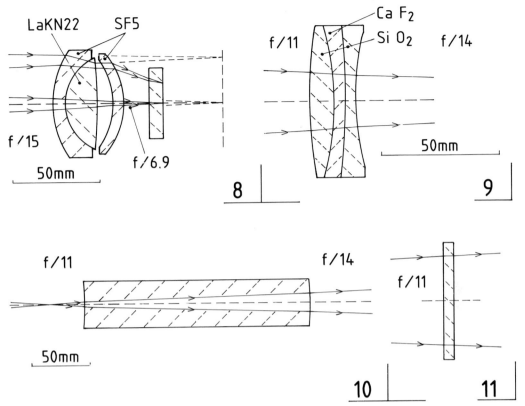

Figures: **8.** The f/15 to f/6.9 reducer of the JKT. **9.** An f/11 to f/14 converter used with refocus. **10.** A doubly parfocal "focal modifier" in one piece of silica. **11.** An afocal, nearly plane-parallel window or colour filter. (The rays in Figures 9, 10 and 11 correspond to the axial field point.)

are well within a specification of about two TV lines (70 μm), and fall to much less than one line near the centre.

An example of the use of a negative lens in a converging beam is provided by a design for a lens which would provide an f/14 image at the Nasmyth or Cassegrain stations of the WHT. The bare mirrors of the telescope provide an f/11 image: the lens shown in Figure 9 reduces the convergence angle and the f/11 and f/14 images are parfocal. This is obtained by designing the lens along with a refocus of the telescope by means of its secondary mirror. The spherical aberration due to the defocus is reduced about 20 per cent by the lens. Fused silica outer elements and a calcium fluoride core are used in this example for the ultraviolet potential, and the lens has good chromatic properties. The f/14 field in this example is 43 mm (2.5 arcmin) in diameter and the image size is within about 0.15 arcsec over the 300–1000 nm range.

Examples of type 4. A "doubly parfocal" lens is shown in Figure 10. This is a "focal modifier rod", consisting of a single rod of fused silica. The rays tend to converge at the front surface and diverge at the second surface, even though these are concave and convex respectively. It is therefore analogous to Figure 4b. This rod provides both a parfocal slit image at the changed f-number, and a parfocal pupil. Longitudinal chromatic aberration is corrected, as the aberration due to the positive power is negated by the effect of the thick slab. There is a some lateral colour which is probably insignificant in a spectrograph, and which is correctable by cementing additional lens elements to the ends. This is designed by the author but no examples have been produced yet.

On the Multiple Mirror Telescope, a single block of silica was used to convert the f/31.6 Cassegrain beam provided by the telescope to f/9 in order to shorten the collimator of the spectrograph. That device was not parfocal for the slit image, although it was for the pupil [Angel et al., 1979].

A "spin-off" from the rod device is a modified plane window or colour filter, a very short rod. Figure 11 shows this device. In this case, the first surface is flat and the second is very slightly convex, such that the position of focus is not affected by the window. The lens power can be extremely small if the window is not close to the focus. Compared to a plane-parallel window in an f/11 beam, the modified window has various advantages. It is parfocal – the axial image does not move when the window is inserted in a certain position. (It is not parfocal as regards the pupil.) It can cause less longitudinal chromatic aberration than a plane window; and the ghost image due to double reflection within the window is further out of focus than that from a plane parallel window.

Examples of type 5. The use of a parfocal relayed image implies that the telescope gives direct images on two different plate scales on a fixed detector, by means of inserting suitable lenses. In practice a focussing mechanism which would be needed in any event is used to take up small differences in the position of focus when lenses are interchanged. These ideas are used extensively in the Acquisition and Guider box [Ellis et al., 1990] for the WHT. The detector (25 mm in diameter) is on a motorised mount, but for only fine focus adjustment, which is applied automatically. A choice of four different images is available on the one detector. These comprise: (i) the direct f/11 Cassegrain image; (ii) the same reduced to f/4.1; (iii) a view of the field reflected from the slit at f/11; and (iv) a view of the field reflected from the slit, but reduced to f/4.1.

A schematic layout of the optics providing these functions is given in Figure 12. The switch to a view of the reflection from the slit is performed by the insertion of a doublet lens and mirror which are mounted on a single slide mechanism. The (separated) doublet lens provides a re-image at unit magnification. Normally the astigmatism of the doublet would be significant, but as the lens is distant from the exit pupil of the Cassegrain telescope, the astigmatism corresponding to a tolerable residual spherical aberration can be made to cancel the normal astigmatism of the

66

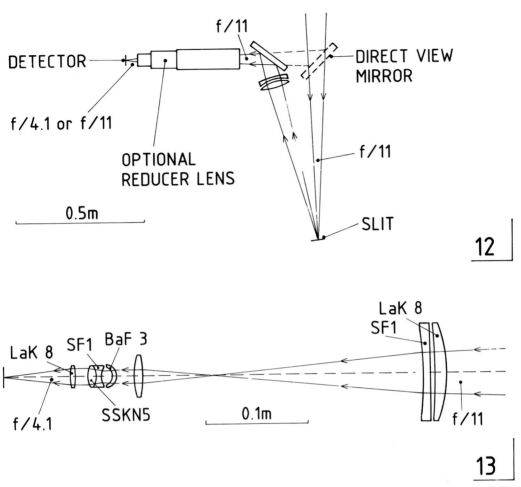

Figures: **12.** Layout of the re-imaging optics in the aquisition and guider unit of the WHT. The two flat mirrors, the doublet lens and reducer lens are all retractable. **13.** The parfocal, direct-view reducer lens in the system of Figure 12.

doublet. The change from f/11 to f/4.1 is achieved by the optional insertion of a parfocal reducer. Its position is shown in Figure 12 and details are in Figure 13. Different reducers are used for the direct view and the view off the slit, because the doublet lens moves the pupil. The re-imaging section of each reducer is a variant of the double Gauss design. The long paths required in Figure 12 tend to increase aberrations, but the various systems perform to a specification of a total image spread of about two TV lines, or $70\,\mu$m total image spread, which corresponds to 0.3 arcsec at f/11 or 0.8 at f/4.1, using a colour filter. The spread of intensity in white light, at the edge of the field, is typically $150\,\mu$m.

Examples of type 6. The "post-focus relay" lens is illustrated by the lens shown in Figure 14. It gives a faster focus and hence a wider field of view on an acquisition TV camera. The field lens in this example is a moulded Fresnel lens and the relay section is a commercial camera lens. One advantage of the Fresnel lens as a field lens is its lack of Petzval field curvature. However, the outer grooves in such a lens may be visible in the image, despite attempts to defocus them, which may be limited by aberrations. The aliasing of the grooves with CCD pixels can give rise to distracting cruciform patterns in background light, but this needs a trial with the real illumination. The relay section also requires experiments if the conjugate distances are outside the focussing range intended for the lens. A system due to I. S. Glass [Glass, 1979], with a 2:1 reduction, is very satisfactory: its TV camera reaches magnitude 20 on a 1.88-m telescope. The "Peoples Photometer" of the Isaac Newton Group telescopes (La Palma) also contains a post-focus relay lens for field acquisition: that example consists of a Cooke-type triplet (not illustrated here) designed to work along with with a cemented doublet field lens.

The 1.88-m telescope at the South African Astronomical Observatory has an f/18 Cassegrain focus. For direct imaging on a CCD, the scale is too large and the lens system shown in Figure 15, designed by the author, converts the f/18 focus to f/6.6. The doublet form of the field lens helps the lens internally by reducing aberrations of the pupil, which is formed within the small relay section. The latter is reminiscent of the historical Zeiss Tessar, with only three components (one cemented). It incorporates special glass types to provide for the range of wavelengths. Aberrations are within about 0.5 arcsec total spread over a field of 18 mm diameter (f/6.6) at 365 to 1014 nm using the whole spectral range, and of course are smaller if a colour filter is used. Observations made with this converter include Halley's Comet [Cosmovici et al., 1988].

The double-Gauss lens type is in principle attractive for the relay section of such systems, but designs tend to include glasses of higher refractive index. That is helpful to the optical designer in some respects but may cut off ultraviolet wavelengths. A preliminary design (not reproduced here) indicates that glasses of relatively low refractive index, around 1.55 or less, may be formed into a double-Gauss for a relative apertures of at least f/4.

Other examples of this type of converter include the polarimeter used by a group from Durham University on a large number of telescopes [Scarrott et al., 1983]. Also, R. F. Griffin and the author recently designed a reducing relay lens for the Cambridge 36-in telescope, converting approximately f/28 to f/14.5. The field lens in that case is a plano-convex element which is cemented to a prism, which acts as one of the coudé reflectors. The lens is cemented to the exit face of the prism; if it was applied to the entrance face, the angles of incidence in the prism would be incorrect for total internal reflection. The relay section is a cemented triplet, embodying a fluoride glass which reduces secondary spectrum.

Examples of type 7. A focal length converter incorporating a collimated space forms the basis of many analytical instruments. Figure 16 shows lenses used in the

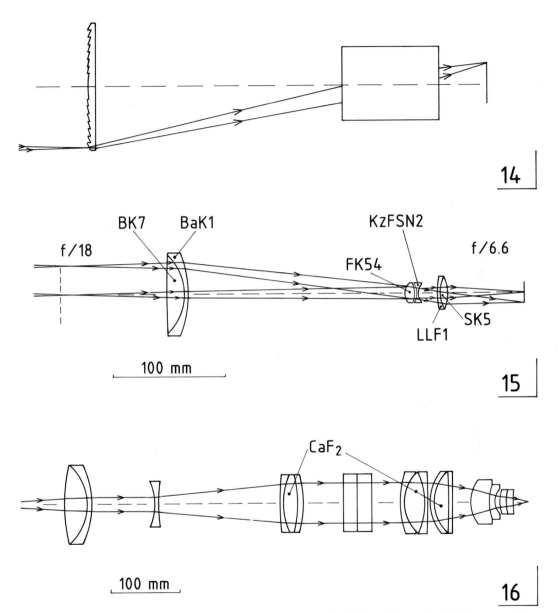

Figures: **14.** A re-imaging system using (schematically) a Fresnel field lens and a commercial relay lens. **15.** A re-imaging system for use on the 1.88-metre telescope at SAAO. **16.** TAURUS II with the f/2 camera on the AAT.

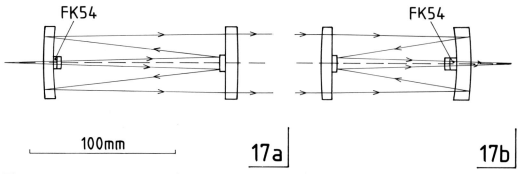

FK54

FK54

100mm

17a

17b

Figures: **17a**. A compact f/15 collimator for use in a re-imaging system with an accessible pupil for experimental optics. **17b**. The f/11 camera. Pupil imagery is such that the system is reversible. The unlabelled lens materials are silica.

TAURUS 2 interferometer [Unger *et al.*, 1990]. They comprise dipotric collimator and camera systems, which were designed jointly by C. F. W. Harmer and the present author. A Fabry–Perot etalon is positioned at a 60 mm pupil image in the collimated space. The system illustrated converts the f/8 Ritchey–Chretién focus of the AAT to f/2: on the WHT, the collimator is adapted to the f/11 Cassegrain focus but the camera is similar. An f/4 camera is also provided on the WHT. Aberrations lead to a total image spread of one arc second in the f/2 camera and one-half arc second in the f/4 camera, over 365 to 1014 nm simultaneously. TAURUS 2 has a field of view on the sky of nine arc minutes at f/2. Further details of these lenses are not given here as they are relatively complicated and it is hoped to publish a full optical description elsewhere.

For image-sharpening experiments on the INT, a system providing collimator, pupil and camera was specified for a field of view of 30 arcsec only, but with long focal lengths. Figure 17a shows the catadioptric f/15 collimator (40 mm aperture). All the surfaces are spherical: the correction of spherical aberration is provided by a Mangin back-surface mirror. The convex mirror is a spherical, front-surface mirror and is mounted at the centre of a window (itself a weak positive lens). Thus there are only two components to mount mechanically. Instead of using a hole through the concave mirror, there is a hole in the coating only and the design uses the silica of the back-surface mirror as part of a lens. This lens has a rather unusual function – it improves aberrations in the pupil. This is to provide for masks and other devices being used in the pupil plane, in this case giving a sharp correspondence between the mask and the primary mirror of the telescope. (Alternatively, an image of a high atmospheric layer is formed at an adaptive mirror. In that case the aberrations in the image of the high altitude layer should be considered.) Figure 17b shows a second system to a similar design but operating at f/11. The two when used together form a re-imaging system. The pupil plane is at the centre of a path of 956 mm between the f/11 and f/15 foci, providing a large working space on an optical-bench arrangement. The whole system can be reversed, whereupon it is matched to the f/11 Cassegrain focus and exit pupil of the WHT, as suggested by J. Noordam. A selection of ordinary microscope objectives accompanies these systems and can be used to enlarge the image onto the final detector. The performance of these systems is diffraction-limited.

References

[Angel et al., 1979] Angel J.R.P., Hilliard R.L., Weymann R.J., *An optical spectrograph for the MMT. The MMT and the future of ground-based astronomy, Smithsonian Astrophysical Observatory, Special Report 385, pp. 87–117, 1979.*

[Bingham, 1978] Bingham R.G., *A two-mirror focal length reducer and field corrector for prime foci. Optical and infrared telescopes for the 1990's, Kitt Peak National Observatory, ed. A. Hewitt (1980), pp. 965–974.*

[Bingham, 1979] Bingham R.G., *Grating spectrometers and spectrographs re-examined. Q. Jl. R. astr. Soc. Vol. 20 (1979), pp. 395–421, particularly pp. 410 and 413.*

[Cosmovici et al., 1988] Cosmovici C.B., Schwarz G., Ip W.-H., Mack P., *Gas and dust jets in the inner coma of comet Halley. Nature, Vol. 332 (1988), pp. 705–709; also ESA SP-250 (1986), Vol. II, pp. 375–379, and ESA SP-278 (1987), pp. 195–207.*

[Courtès, 1972] Courtès G., *Spiral structure and kinematics of the Galaxy from a study of the HII regions. Vistas in Astronomy, Vol. 14 (1972), pp. 81–161.*

[Diego & Walker, 1985] Diego F., Walker D.D., *Increasing the throughput of astronomical spectrographs by overfilling the dispersing element. Mon. Not. R. astr. Soc. Vol. 217 (1985), pp. 347–354.*

[Ellis et al., 1990] Ellis P.A., Bingham R.G., Worswick S.P, *Acquisition and guider unit for the Cassegrain focus of the 4.2m William Herschel telescope. Proc. Soc. Photo-Opt. Instr. Engrs. Vol. 1235 (1990), pp. 777–785.*

[Glass, 1979] Glass I.S., *A simple focal reducer using a Fresnel lens. Mon. Not. astr. Soc. S. Africa Vol. 38 (1979), pp. 38–39; see also Vol. 44 (1985), pp. 45–47.*

[Sánchez & Wall, 1991] Sánchez F., Wall, J.V., *The Canarian Observatories. Astro. Lett. and Communications, Vol. 28 (1991), pp. 47–67.*

[Scarrott et al., 1983] Scarrott S.M., Warren-Smith R.F., Pallister W.S., Axon D.J., Bingham R.G., *Electronographic polarimetry: the Durham polarimeter Mon. Not. R. astr. Soc. Vol. 204 (1983), pp. 1163–1177.*

[Unger *et al.*, 1990] Unger S.W., Taylor K., Pedlar A., Ghataure H.S., Penston M.V., Robinson R., *The nature of the high-velocity gas in NGC 1275: first results with TAURUS-2 on the William Herschel telescope. Mon. Not. R. astr. Soc. Vol. 242 (1990), pp. 33p–39p.*

[Welford, 1986] Welford W.T., *Aberrations of optical systems. Adam Hilger, 1986.*

7

Recent CCD Developments at RGO and ING, La Palma

Paul R. Jorden*, A. P. Oates* and A. S. Ahmad *

Abstract

All three telescopes of the Isaac Newton Group have been equipped with small-format CCD cameras since they were first commissioned (1984 onwards). Within the last two years these have been upgraded to utilise larger arrays (from 770×1152 to 2186×1152 pixels). The performance improvements of these larger detectors is discussed.

The 'standard' operating temperature of our CCDs is 150 K; we have recently quantified the optical sensitivity of a CCD at a range of higher (and lower) temperatures. By operating above our usual temperature, significant improvements in quantum efficiency, at wavelengths around 1 μm, can be achieved.

Other current developments in CCD systems at RGO and ING are presented.

1. Introduction

At RGO we have developed and commissioned a substantial suite of CCD systems within the last decade; all of our Isaac Newton Group (ING) telescopes use CCD sensors. The Charge Coupled Device (CCD) has seen common use at major observatories for many years, so scientific advantages and technical characteristics will not be detailed here. Further details may be found in other publications [Jacoby, 1990; McLean, 1989].

This paper briefly reviews these operational detectors and summarises their characteristics and use at a variety of direct-imaging focal stations and spectroscopic instruments. Much initial development and device-testing has been undertaken at RGO (Herstmonceux and Cambridge), and some results and future plans are presented.

The material presented here is based on the September 1991 Herstmonceux Conference, with a few more recent updates prior to publication of these proceedings.

---

*Royal Greenwich Observatory, Madingley Road, Cambridge CB3 0EZ, UK.

2. CCDs at the Isaac Newton group of telescopes

Table 1. ING Telescope Foci and Instruments*

Focal station	Tel/instrument field-of-view		Image scale (at detector)	CCD FOV		
	(arcmin)	(mm)	(arcsec/mm)	CCD	(arcmin)	(mm)
WHT f/2.8 Prime	40/14	140/49	17.6	TEK1:	7.2	24×24
WHT f/11 aux	15/4.5	200/60	4.5	TEK1:	1.8	24×24
WHT ISIS	$\lambda \times 4$	35×17	14.9	EEV3:	$\lambda \times 6$	28×26
WHT UES	-	38×16	15.5	EEV8:	-	49×26
WHT LDSS f/2	11.4	27	25	TEK1:	10×10	24×24
WHT Taurus f/4	9	27	12.4	TEK1:	5×5	24×24
INT f/3.3 Prime	40/12.5	96/30	24.7	EEV5:	11.5×10.7	28×26
INT IDS/235cam	$\lambda \times 4$	30×8	33	EEV5:	$\lambda \times 13$	28×26
JKT f/15 Cass	34/9	150/40	13.8	EEV7:	6.4×5.6	28×26

*One CCD head is selected as an example for each focus; alternatives may be used in many cases.

2.1. Focal stations

The ING telescopes on La Palma (4.2-m WHT, 2.5-m INT, and 1.0-m JKT) have all been equipped with CCDs since original commissioning. This situation has not remained static, and most of the earlier small-format CCDs have now been replaced by larger-format, lower-noise arrays.

Since we have a variety of instrument configurations (on all telescopes), all CCD-heads have been built to the same mechanical design with identical focal-plane positions, to facilitate exchanges between instruments.

Table 1 shows most of the major telescope and instrument foci where CCDs can be used. A few instruments have embedded CCDs (FOS1, FOS2 and WYFFOS), and are not described here. Full details of the telescope facilities can be found in the *ING Observer's Guide* [ING Observers Guide, 1988]; some of the CCD instruments are described elsewhere [Jorden *et al.*, 1986].

The second column of the table, shows the field-of-view provided by the telescope focus, or instrument. At direct foci, the full telescope FOV is often reduced by instrument platform constraints (e.g. filter sizes etc.); these are indicated by the two figures given. On the spectrographs, the detector field encompasses a spatial and spectral component; the spectral component (represented by λ in the table) is not defined here, since it depends on grating configuration.

It can be seen that in many cases the detector field-of-view does not completely fill the available focal plane area. Most of our spectrographs were designed for

Table 2. Currently available CCDs at ING Telescopes

Detector	CCD type	Format	Telescope	Uses
GEC3.1	P8603	385×578	INT (JKT)	IDS (Prime, photometer)
FOS1	P8603	385×578	INT	Fixed function spectroscopy
RCA2	SID501	320×512	INT (JKT)	Prime (IDS, photometer)
GEC6	P8603	385×578	JKT (INT)	Photometer (IDS)
GEC5	P8603	385×578	WHT	Aux imaging, ISIS, GHRIL
FOS2	P8603	385×578	WHT	Fixed function spectroscopy
EEV3	CCD-05-30	1242×1152	WHT	ISIS
EEV4	CCD-05-20	770×1152	WHT	Aux imaging, GHRIL, ISIS
EEV5	CCD-05-30	1242×1152	INT (JKT)	IDS, Prime
EEV6	CCD-05-30	1242×1152	WHT	ISIS
EEV7	CCD-05-30	1242×1152	JKT (INT)	Photometer, imaging
TEK1	TK1024	1024×1024	WHT	LDSS, ISIS, imaging, UES
GEC7	EEV-02-06	~335×520	INT (JKT)	Prime, IDS (photometer)

use with the IPCS (Image Photon Counting System [Boksenberg & Burgess, 1972]) which has a 40 mm diameter sensitive photocathode. The original small-format CCD arrays only captured part of the available spectrograph field, but have still been widely used because of their high sensitivity. The more recent large-format heads (EEV, TEK) have increased our focal-plane photon capture by 6–10 times.

In order to illustrate the ranges of use of CCDs, Table 2 presents a list of currently available detector heads at the ING. The primary use of each head is given, although most are used interchangeably with heads from other focal stations.

2.2.   Detector properties

The CCD has been described as the ideal detector; this is not yet true! Various manufacturers have continued to develop the device and we have attempted to use the best available sensors. This has led to a suite of camera-heads, providing a range of detector options on all telescopes.

Table 3 summarises the properties of the main types of array that we have used on our ING telescopes. The quantum efficiency (QE) is only indicated at two selected wavelengths; QE over the whole spectral range is discussed below.

Table 3. Parameters of operational ING detectors

Detector type:	GEC P8603	RCA SID501	EEV-05-30	TEK-1024
Format	385×578	320×512	1242×1152	1024×1024
Pixel size (μm)	22×22	30×30	22.5×22.5	24×24
Image area (mm)	8.5×12.7	9.6×15.3	28×26	24×24
read noise (e^-)	6-10	60	3-5	5-10
QE (%) 650 nm :	55	75	50	75
400 nm :	15 (coated)	40	15 (coated)	55

Other devices exist and are used elsewhere; our choice of sensor has been determined by performance, availability, operational support and sometimes cost. The performance of CCDs has certainly improved considerably over the last ten years or so; more details may be found in these references: [Janesick *et al.*, 1984; Jorden, 1990a].

As can be seen in Table 2, we have used many GEC and EEV sensors. The current EEV CCD-05-xx series of devices offers very competitive performance; only thinned (higher QE) chips from, for example, Tektronix, offer significantly improved performance for some projects. The first large-format sensor that we used on our telescopes was an EEV 05-20 (or P88200) device. This has been described elsewhere [Bailey & Pool, 1990; Jorden, 1990b], and its performance is typical of all devices of this type.

2.3.   Scientific applications

The ING telescopes all have 'common-user' instruments; this means that they are in routine use by visiting astronomers who may not be familiar with them. In addition, they have to be supported by technical staff who generally were not involved in their development or construction. We have the additional requirement that CCD heads must be shared and interchanged between instruments.

The previous tables have indicated the types of telescope and instrument combinations that we have available. Not all applications can be described here, but we shall highlight some of them, with indications of the constraints on CCD-type.

2.3.1.   Direct imaging

Available at WHT (GHRIL, Auxiliary Cassegrain, Prime and re-imaging instruments), INT (Prime), and JKT (Cassegrain).

This mode is used mainly for photometry, morphology, faint identifications and astrometry. Usually, exposures are made with broad-band filters, and the dominant noise is due to Poisson noise from the sky background. In these cases, the detector readout noise may not be crucial, and a high quantum efficiency is more important. A large focal-plane detector area is also most desirable for this type of observation. Other important factors are: uniformity of response, stability (of spatial sampling, and signal gain), and wide dynamic range.

2.3.2.   Medium-resolution spectroscopy

Available at WHT (ISIS, FOS2), INT (IDS).

For this type of work, a good QE over a wide spectral range, with low readout-noise is important. For rectangular devices, the major axis is normally aligned to capture the maximum wavelength range; these CCDs also have sufficient width in the spatial direction for observations of most extended objects. A wide dynamic range is particularly useful for (arc) calibration spectra, since line intensities may vary by several orders of magnitude. The ability to read out selected windows reduces read-out time, and economises on data storage.

2.3.3.   High resolution spectroscopy (WHT UES)

In this case, sky background levels are usually negligible, and source signal levels are low. Hence, a detector with the lowest possible read-out noise is very important; it also helps if this is coupled with a high QE over a wide wavelength-range. The cross-dispersed echelle is particularly efficient with a large-area detector, which enables it to record all the available spectral orders.

3.   Choice of detector

3.1.   General characteristics

There is a diverse range of CCDs available for scientific use. This remark should be qualified by the comment that there is no single CCD that ideally matches all our requirements. The ideal case would be to have a complete set of detectors, to suit all needs; but this would be expensive to procure and demanding to support. We inevitably adopt the compromise position of having a limited set of arrays covering most of our requirements.

At the time of our telescope commissioning (in the mid-1980s), only small-format arrays were available, so the lowest-noise versions that were available (GEC P8603s) were selected. Now (1990–1992), larger-format arrays are available, and we have been installing these at all focal stations. These newer sensors also have the benefit of lower noise (EEV 05-xx series).

Fig. 1 shows a selection of currently available CCDs, indicating the range of formats available. Each has its own merits; performance, ease of use, cost and availability vary quite widely.

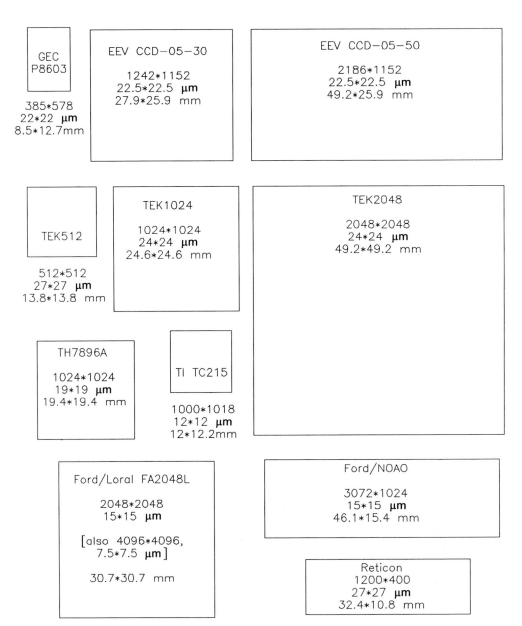

A selection of current scientific CCDs (to scale) [PRJ 1992]

Figure 1. Examples of currently available CCDs.

At RGO we have invested much time in characterising various new devices. Initially, we choose prototypes to evaluate for potential use, and then often purchase and test further grade-1 devices for telescope use. All detectors are optimised for the lowest practical readout-noise, and best charge-transfer performance. Linearity, dark-current and other important characteristics are measured. We also use our photometric laboratory to calibrate the QE of these arrays at cryogenic temperatures. Some recent results are presented in the next section.

3.2.   Quantum efficiency *vs* wavelength and temperature

Some manufacturers of CCDs measure spectral response, but often only at room temperature. Furthermore, although they are interested in the general responsivity, they do not necessarily have the precision of measurement that we require.

We characterise all of our CCDs before they are commissioned at the telescopes; this is carried out using our 'Jelley' photometric system. Fig. 2 illustrates the components of this calibration system, which is described fully in a technical note [Jelley, 1983]. We also monitor CCD performance with portable Beta-light standards, which are available at the ING telescopes [Jorden *et al.*, 1992].

Fig. 3 shows plots of QE versus wavelength for the CCDs that we currently use on our ING telescopes. These 'standard' camera-heads all operate at low temperatures (140–185 K), giving negligible dark current, and hence permitting long exposures.

In addition, we have recently characterised the spectral response of a CCD over a range of temperatures. It is known that the absorption efficiency of silicon changes with temperature (see, for example [CCD data-book, 1987]), in particular a reduction in near-infrared response as the device is cooled. For this investigation a standard, small-format, dye-coated P8603 CCD was evaluated at a range of temperatures from 110 to 210 K. Fig. 4 illustrates the spectral response curves that were obtained, at the centre and two extremes of our measured temperature range. It can be seen that the response increases with temperature at longer wavelengths, as expected. Fig. 5 highlights this effect by plotting the relative QE for each wavelength versus temperature. For example, at the longest wavelength the response increases threefold by raising the temperature from 150 to 210 K. However, the use of a higher temperature does result in a higher dark current, and so readout noise can increase. These effects are discussed in the next section.

3.3.   Dark current and readout noise

Dark current within the CCD is a strong function of temperature, following a diode law, with an exponential dependence on temperature. We normally choose our operating temperature so that for an exposure time of one hour the total integrated dark signal is very small. As temperature increases this is no longer true, and

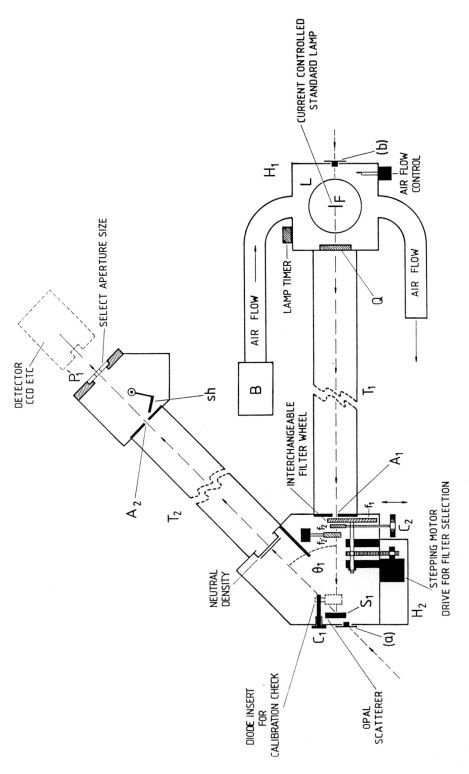

DETECTIVE QUANTUM EFFICIENCY MEASUREMENTS

Figure 2: The RGO spectral response measurement system.

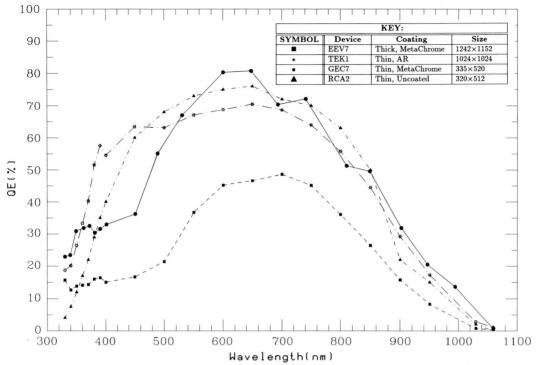

Figure 3. Quantum efficiency versus wavelength for current ING CCDs.

the dark signal contributes an increasing amount to the total noise of the readout process.

It is not intended in this paper to discuss all the sources of noise within a CCD. The important point is that the dark signal (N electrons) has an associated Poisson noise $\sqrt{N}$; this may exceed the normal (random) readout noise, and therefore be the dominant source of noise in the measured signal. The basic read-out noise can also be a function of temperature; for the chip tested our read-noise was about 11 e^- rms everywhere, except at the lowest temperatures where it increased somewhat.

Table 4 shows the measured dark-signal, associated Poisson-noise, and total measured noise as a function of temperature. Note that at 110 and 130 K the dark current was too low to measure; at 150 K the precision is low.

The conclusion is that an elevated temperature can increase the long-wave spectral response, but also increases the dark-current noise. From our tests, it is clear that for observations around 1 μm the CCD (that we tested) should be operated at an elevated temperature of say 190 K.

The values presented here are specific to the chip tested, but the general conclusion is still valid. The exact point at which it is worthwhile elevating the temperature depends on several parameters: basic readout-noise of the CCD, wavelength to

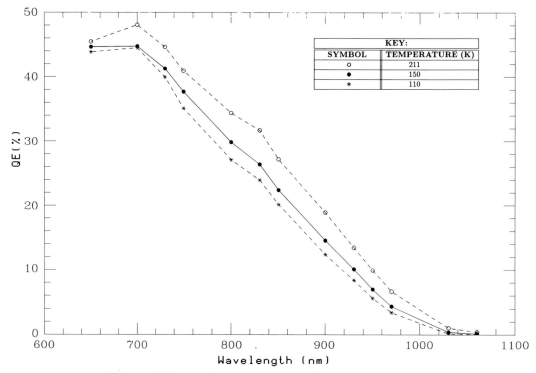

Figure 4. Spectral response at various temperatures.

Table 4. Dark signal and total readout noise versus temperature

Temperature (K)	Dark signal (in 2000 s) (e⁻/pixel)	Poisson noise (e⁻)	Readout noise (e⁻ rms)	Total noise (e⁻)
110	-	-	26	26
130	-	-	15	15
150	3	1.7	11	11
170	6	2.5	14	14
190	60	7.8	13	15
210	3060	55	13	57
230	51600	227	15	227

be measured, exposure time to be used. The final signal/noise can be increased, provided that the QE increases in greater proportion than the dark-current noise, for a given wavelength and temperature.

Table 5 summarises these results by showing an example of the S/N improvement for one set of conditions. A more extensive description of these results can be found in the reference [Ahmad *et al.*, 1993].

82

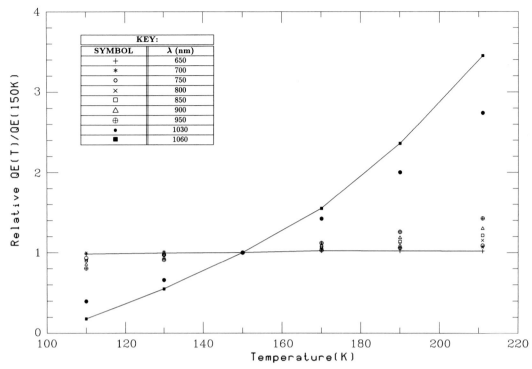

Figure 5. Relative QE versus temperature.

Table 5. Example of S/N enhancement at 1060 nm, using a 2000-s exposure, at λ = 1060 nm

Temperature:	150K	190K
Dark Current, N (e⁻)	3	60
Readout Noise, σ_r (e⁻ rms)	11	13
Total Noise, $\sigma_T = \sqrt{N + \sigma_r^2}$	11	15
Quantum Efficiency, QE (%)	0.11	0.26
'S/N'= $QE/\sigma_T \times 10^4$	1.0	1.7

Hence, Signal/Noise at 190 K (*cf* 150 K) is a factor of 1.7 better.

4. Future directions

There is a variety of developments taking place at RGO and elsewhere to improve the usefulness of CCDs for astronomy. It is impossible to review all topics in detail here, but we shall attempt to highlight and summarise some of them.

83

Pixel sizes are getting smaller. Average sizes are decreasing from 10–30 μm to 5–20 μm. This is mainly led by the fact that the manufacturing yield of CCD arrays is better for small areas, and so for a given pixel count a small size is best. This is a mixed blessing. For some instruments, we can achieve a better spatial resolution, on others we shall have to suffer a smaller total field size.

Larger array formats are being manufactured; the largest to-date is the 4096-format device, manufactured by Ford/Loral [Janesick et al., 1989]. However, the maximum practical array size is of order 100 mm (diagonal), limited by the current size of industry-standard silicon wafers. Although such a device would fit on one wafer, no manufacturer is very close to producing such a large chip successfully.

It may be noted (from Table 1) that direct-imaging telescope foci provide an available area that considerably exceeds the CCD array size. The only practical solution to this problem is the use of a mosaic of CCDs. Such a system is currently under development for the INT Prime Focus as a survey facility (jointly funded with C. Pennypacker, Berkeley). A mosaic of 2×2 (2048) CCDs has also been described recently in [Geary et al., 1991].

The readout noise of CCDs is decreasing. Our early (GEC) CCDs had a noise figure of about 6–8 electrons rms; our current (EEV) sensors operate with 3–4 electrons rms noise. Devices with even lower figures are becoming available, with sub-electron performance obtainable in some circumstances [Janesick et al., 1989]. This allows us to achieve better sensitivity (at low light levels), as well as a greater dynamic range.

There are several other areas of architectural and design changes that benefit users of such devices. The 'notch' design can much reduce the CCD sensitivity to defects, and give improved charge-transfer efficiency. The Multi-phase-pinned (MPP) device yields a much lower dark current, which can allow long exposures at higher temperature. Multi-output devices are now commonplace; four outputs (one in each corner) are now routine for large arrays, in order to minimise readout times through parallel data-collection techniques.

An important field of very active research is the subject of thinning, and high QE. Various manufacturers have achieved good results, but only with great difficulty (and expense). Several US astronomers and manufacturers are working to develop better thinned CCDs; similarly, we are also collaborating with EEV in the UK. Continued development of thinning technique and anti-reflection coating should lead to larger, high-efficiency CCDs with good uniformity and stability.

In general, suggestions from scientific users such as ourselves provide important feedback for manufacturers of CCDs. We measure device characteristics with some care, chiefly spectral response and readout noise, and this data is conveyed to suppliers like EEV. These results often enable modified devices to be produced with superior performance, to the mutual benefit of manufacturers and the astronomical community.

Acknowledgements

Much of the data presented here could not have been taken without the support of P. Terry, in our photometry laboratory. R. Lawrence helped to upgrade our microprocessor-based photometry system. Many other RGO staff have contributed to the successful construction and operation of our CCD systems. EEV and GEC are acknowledged for many useful technical discussions.

References

[Ahmad *et al.*, 1993] Ahmad A.S., Jorden P.R., Oates A.P., Terry P., *CCD quantum efficiency characterisation and enhancement. RGO Technical Note, 1993.*

[Bailey & Pool, 1990] Bailey P., Pool P., *The P88000 series of large area CCDs for scientific imaging applications. Proc. SPIE, Vol. 1191 (1990), p. 23.*

[Boksenberg & Burgess, 1972] Boksenberg A., Burgess D.E., *The Image Photon Counting System. Adv. E. E. P., Vol. 33B (1972), p. 835.*

[CCD data-book, 1987] *CCD-Imaging-III data-book, EEV: Chelmsford, Essex, 1987.*

[Geary *et al.*, 1991] Geary J.C., Luppino G.A., Bredthauer R., Hlivak R.J., Robinson L., *CCDs and Solid State Sensors II. Proc. SPIE, (1991), in press.*

[ING Observers Guide, 1988] *ING Observers Guide, eds Unger S.W. et al., RGO, 1988.*

[Jacoby, 1990] *CCDs in Astronomy, Vol. 8. ed. Jacoby G., Astron. Soc. Pacif. (see various papers by Janesick and others), 1990.*

[Janesick *et al.*, 1989] Janesick J., Elliot E., Dinzigan A., Bredthauer R., Chandler C., Westphal J., Gunn J., *New advancements in CCD technology. in CCDs in Astronomy, ed. Jacoby G., Astron. Soc. Pacif., 1989.*

[Janesick *et al.*, 1984] Janesick J., Elliott T., Collins S., Marsh H., Blouke M., Freeman J., *The future scientific CCD. Proc. SPIE, Vol. 501 (1984), p. 2.*

[Jelley, 1983] Jelley J.V., *Primary and secondary light standards at RGO. RGO Technical Report, 1983.*

[Jorden *et al.*, 1986] Jorden P.R., Thorne D.J., Waltham N.R., *Four CCD instruments at the La Palma Observatory. in Proc. ESO/OHP workshop on "Optimisation of the use of CCDs in astronomy", 1986, p. 13.*

[Jorden, 1990a] Jorden P.R., *CCDs for the 1990s. in New Windows on the Universe, eds Vazquez M. & Sanchez F., Cambridge University Press, 1990, p. 465.*

[Jorden, 1990b] Jorden P.R., *An EEV large-format CCD camera on the WHT ISIS spectrograph. Proc. SPIE "Instrumentation in Astronomy VII", Vol. 1235 (1990), p. 790.*

[Jorden *et al.*, 1992] Jorden P.R., Terry P., Oates A.P., *Beta-light sources for CCD response measurements. RGO Technical Note 86, 1992.*

[McLean, 1989] McLean I.S., *Electronic and computer-aided astronomy. Ellis Horwood, 1989.*

III New and future instrumentation

8
Optical Instruments: Where We Go From Here

David Walker *

Abstract

This paper presents a 'broad-brush' review of some aspects of the development of astronomical instrumentation, with the aim of identifying and exploring an underlying philosophy which can help to guide us in the future. Historical cases are cited, leading to examples of more specific research contributions of the Optical Science Laboratory at UCL.

1. Introduction

It has proved instructive to attempt to tie together a number of strands in diverse areas of optical instrumentation for astronomy. The aim was to identify a coherent underlying philosophy which may usefully seed ideas for future work. The philosophy is that of 'mosaicing'. This can be perceived as a method of splitting a difficult problem into manageable sections. This impacts on the cost, risk and timeliness of instrumentation development, and indeed opens the way to achieve what would otherwise be impossible.

It is remarkable how widespread is mosaicing; indeed in some cases it may hardly be recognised for what it is. Alerted to the widespread benefits of the technique, we may usefully look in future instruments for new openings to apply it.

2. A historical case

In considering future possibilities, lessons can be learnt from past history. In particular, it is illuminating to compare the development of two detectors – the CCD and IPCS. The former was invented at the Bell Laboratories in 1970s [Boyle & Smith, 1970]. By 1973 there were laboratory 320 by 512 pixel optical imagers, but according to McLean [1989], the first experimental astronomical CCD image (of planetary limb-brightening) was not recorded until 1976. The first CCD camera at KPNO came in 1979. In the same year, the UK Steering Committee on Telescopes ('SCOT') recognised [Morgan, 1979] the UK's need for CCDs and noted that 'the very serious, perhaps overriding problem should be mentioned, namely availability'.

---

*Department of Physics and Astronomy, University College London, Gower Street, London WC1E 6BT, UK.

The original RGO camera was developed shortly afterwards, followed by the UCL camera [Walker *et al.*, 1985] (commissioned late 1982) for the 1-m telescope at SAAO.

These cameras of course still used the small-format devices. In 1985, the Tektronix company *announced* the 2048 square CCD, which was successfully produced only recently, and then in only the unthinned state. EEV produced the 1242×1152 device in about 1989.

Remarkably, the development of the IPCS commenced at UCL in the *same year* as the CCD i.e. 1970 [Boksenberg, 1971]. However, by 1973 it was already in 1-D astronomical use [Boksenberg & Burgess, 1973], and this led rapidly to a long-term front-rank programme on several major telescopes (in particular the 5-m Hale), in various observational fields such as QSO absorption-line studies. Extended 2-D detection was in regular use by 1979. Even this elapsed period was more due to large-scale memory availability than intrinsic detector development.

The reason for the remarkably fast development of photon counting was undoubtedly the philosophy of using a standard bought-in EMI image intensifier and Plumbicon TV camera. The system thus benefited from the considerable *prior* industrial R&D investment in the most critical front-end components. This is intimately related to the concept of *risk management* in instrumental development, as outlined below.

3. Risk

In any significant instrumentation development programme, there is bound to be some element of risk. An attempt at completely risk-free development will stifle innovation and often result in marginal performance benefits at best.

Risk can take many forms, such as:

1. risk of exceeding budget (How do you cost innovation?)

2. risk of not meeting schedule

3. risk of not meeting detailed performance specification

4. risk of unsatisfactory performance in unexpected (and unspecified) areas

5. risk that performance will degrade with time or use

6. risk of unreliability in service

7. risk that the technology will be overtaken by something better before the development is complete

8. risk of catastrophic failure ('it breaks')

9. risk of ergonomic failure (works fine, but the users don't like it)

The larger the technological steps forward, the larger in general are both the risks and the potential benefits. Risk may surface as a 'soft' ignorance in estimating costs, timescales etc., or through some 'hard' unexpected technological problem which requires significant additional resources for its resolution (if indeed it can be fully resolved). An example of the latter was the very early attempts by EEV to produce large (1500×1500) CCDs, some of which cracked on cooling. Fortunately this problem was resolved in their P88000 series of larger devices. The greater the technological advances required, the longer the development cycle tends to become and, ironically, the more likely is point 7 above to become true! This might be illustrated by image photon counting techniques displacing electronography, and more recently, photon counters being increasingly displaced by CCDs in many (but by no means all) applications.

Attention has already been drawn to the use of standard components in the original IPCS to reduce risk, and speed development. This itself did not involve mosaicing technology. However, it does lead naturally to the notion that mosaicing standard or well-proven technology can improve the ratio of benefit to risk.

In the next sections, various areas of astronomical technology are explored with this philosophy in view.

4.   Mosaicing in telescopes

The jump from 4-m class to 8-m class telescopes has implied an enormous technical challenge in the primary mirror technology, with the attendant element of risk. Two solutions involving mosaicing illustrate two very different solutions – one is obvious, the other more subtle. A third possibility for the future is explored which combines good features of both approaches.

4.1.   The Keck telescope

The first case is well-known – that of the Keck telescope which has 36 actively-controlled hexagonal mirrors which tile a 10-m aperture. Nelson has reviewed the project [Nelson & Mast, 1990] and details of the technology have been extensively published, for example in a series of seven papers in *Proc. SPIE*, Vol. 1236 [1990]. An important lesson we may learn is that the telescope is entering operation years ahead of projects aiming to construct monolithic 8-m class mirrors. This is because the Keck approach has effectively circumvented one of the key areas of serious risk – large light-weight structures in glass-type materials which are effectively 'un-mendable'. The composite mirror achieves stiffness by virtue of the active control system, a technique which in other disciplines takes the name 'smart structures'. Unfortunately, the solution introduced a different technical problem – the figuring of the individual hexagonal segments. Circular blanks were polished and then cut hexagonal, and warping of the figure resulted. The current solution is the provision in the telescope of mechanical 'warping harnesses'.

4.2. Large aluminium mirrors

The second case is less well known. This is mosaicing at an earlier stage of manufacture than the finished blank. In particular, attention is drawn to the work of the Telas organisation in France [Rozelot & Leblanc, 1991]. Telas is an economic grouping of French companies Aerospatiale and Framatome. Telas is developing aluminium mirrors for 8-m class telescopes (and smaller). The project, which has the European 'EUREKA' label, goes under the name 'LAMA' (Large Active Mirrors in Aluminium). The philosophy is to assemble a large mosaiced blank by welding together sections of forged aluminium, using electron-beam techniques in a vacuum. This is a standard industrial technique used, for example, for pressure vessels. The surface would be machined, then hard-nickel coated prior to polishing by conventional methods. A standard evaporated reflecting coating would be applied. We see here again the emergence of well-proven industrial methods as a low-risk solution for astronomy.

The geometry of a LAMA mirror would be thin meniscus, for example compatible with the ESO VLT. Because the density and Young's modulus of aluminium are about the same as for glass, a common design of passive and active support system may be used.

The advantages of aluminium are several. It has high thermal conductivity, and stays in thermal equilibrium with its environment, minimising 'mirror seeing'. It has some ten times the yield stress of glass-type materials, and so is considerably more robust, reducing risk of damage e.g. in handling or by malfunction of the support system. This mosaicing technology lends itself to repair at any stage of its manufacture, unlike glass and glass-ceramic materials. The historical objection has been long-term creep, but this is completely circumvented in the context of active support.

Under contract to ESO, Telas have produced a 1.8-m mirror blank by the above process. This was successfully figured spherical, and no signature from the welded joins was visible in the final interferogram. A sequence of thermal cycling tests confirmed that the figure was retained. ESO's report [Dierickx & Zigman, 1991] was very optimistic regarding the applicability of the technology to 8-m telescopes.

4.3. Other mosaicing geometries

Bingham *et al.* [1990] have explored mosaicing geometries other than hexagonal. The particular emphasis is on implementing the mosaic as a thin, actively-controlled meniscus. This could possibly be in nickel-coated aluminium. In particular, a radially-segmented ('pie-slice') mirror has significant advantages (Fig. 1). By keeping the periphery of the mirror circular, the mirror could be figured in the assembled state on the polishing machine. The polishing process would then reduce to the classical radially-symmetric problem, avoiding the need to produce individual off-axis segments and then handle warpage on cutting. Precautions would be needed

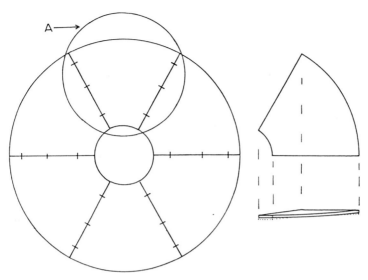

Figure 1. An 8-m 'pie-slice' mirror, and how the segments fit into a 4-m aluminising chamber ('A').

to avoid 'rolling' of the edges of the segments during polishing. This is a current area of study, but the application of controlled forces to the mirror blank along the edges seems a promising approach.

There are advantages of the approach in addition to optical figuring outlined above. For example, in the case of an 8-m telescope, each pie-slice could fit within existing 4-m aluminising facilities, with considerable savings in capital expenditure. Similarly, the blanks could be profiled on a 4-m capacity diamond generating machine. A spare mirror segment could in principle be provided by frequent substitution throughout optical production. In use, inclinations of the mirror segments could be slightly offset to mimic some of the features of an image slicer.

Another potential application is in space where a radially-mosaiced mirror could be launched in a compact folded form, and subsequently deployed. In particular, the incorporation of an active control for the mirror in service reduces the precision with which the deployment mechanism would have to work from optical to engineering tolerances.

5.   Mosaicing in optical instruments

5.1.   Optics in echelle spectrographs

Instruments for 4-m class telescopes have in certain areas approached or even already exceeded the limit of what monolithic optical solutions can provide. This is particularly so in the case of high-resolution spectroscopy. For example, the UCL Echelle Spectrograph (UCLES) [Walker, 1987] employed the largest monolithic echelle grating produced by Milton Roy. Three fused silica prisms in tandem provided cross-dispersion, each being a mosaic of three wedges oiled together. It is believed that these prisms are the largest high-quality silica prisms ever produced.

Transporting the UCLES design to an 8-m telescope would halve both the slit width and order separations in arcsec. Recovering the slit width demands either a larger beam size in the echelle dispersion direction, or a higher blaze angle echelle, both requiring a longer ruled length of echelle. The Lick group building the Keck 'HIRES' echelle [Vogt & Penrod, 1987] opted for an R2.6 special ruling (cf. the standard R2 in UCLES), with three echelles mosaiced end-on-end. Similarly, recovering order separations requires either increased prismatic dispersion, or an enlarged beam in the cross-dispersion direction. For UV transparency to the atmospheric cut-off, fused silica is the only practical choice. This implies the need either for more prisms, or larger prism angles, or double-passed prisms, or an enlarged beam in the cross-dispersion direction, or a combination of the aforementioned possibilities. All of these introduce severe problems in prism size and hence fabrication.

The Keck HIRES avoided the prism-size problem by taking the pragmatic approach of using grating cross-dispersion. Whilst a practical solution, this falls short of the ideal, not least because of the extremely non-uniform order spacings which gratings produce. Other problems include the anomalous polarisation properties of low-order gratings, the presence of a grating blaze profile (leading to much reduced throughput in the corners of the echellogram cf. prisms) and lower overall efficiency.

The above considerations prompted us to explore prism mosaicing possibilities, working from the premise that the material *inside* a prism is costly, serves no useful purpose, and could be dispensed with. This led to a family of solutions (e.g. Fig. 2), based on the 'Christmas tree concept', discussed in more detail elsewhere [Walker et al., 1990]. The 'prismlets' might be blocked up and polished together on a large lap, thus ensuring precisely-matched apex angles.

5.2. Mosaiced CCDs

Proceeding from a 4-m to an 8-m telescope also leads to a problem of detector real estate. Other things being equal, spectrographs for larger telescopes demand the same number of pixels but of increased linear size. Otherwise, the spectrograph cameras become impossibly fast. However, the industrial tendency is to decrease pixel sizes as this improves manufacturing yield. As mentioned at the start of this paper, large CCDs have been slow to reach commercial production. Mosaicing presents a low-risk escape from the dilemma, and opens the way for image formats impossible otherwise within the forseeable future. A specific example is the planned Gemini direct imaging camera [NSF, 1989], the preliminary design of which uses a three by three array of large-format CCDs, which could be Tektronix 2048 by 2048

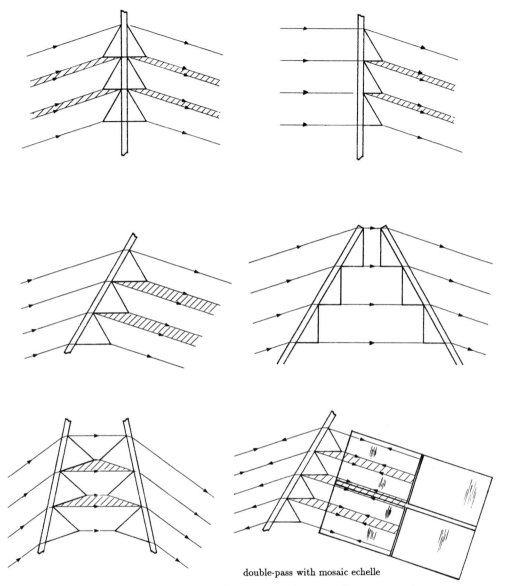

double-pass with mosaic echelle

Figure 2. Possible designs of segmented cross-disperser prisms.

devices or equivalents from some other manufacturer. Another example is the Keck 'LRIS' (Low Resolution Imaging Spectrograph), the camera of which [Epps, 1990] is designed to feed a two by two mosaic of Ford 2048 by 2048 CCDs with 15 μm pixels.

Mosaicing of smaller CCDs also has advantages over the equivalent monolithic solution, such as:

Potential for parallel readout of multiple devices, reducing total readout time.

Possibility of mixing devices e.g. a spectrum format could be optimally recorded with thinned CCDs for the UV/blue, thick for the red, and even IR-arrays for the infrared. This leads on to:

Large image formats can be covered with new-technology devices (e.g. for the IR), possibly years before monolithic devices of the same size become available.

The geometry of the mosaic can be tailored for a specific spectrum format e.g. the trapezoidal shape of an echellogram. This not only maximises the efficiency with which the detector real-estate is used, but more importantly, it improves geometrical throughput when the detector itself is the physical obstruction within an unfolded spectrograph camera.

The single most difficult aspect of mosaicing CCDs is reduction of the butting gaps. The areas of normal commercial CCD packages often exceed the active areas by more than 50%. The observational importance of gaps in the image coverage is somewhat controversial. In some applications even large gaps may hardly be an issue. For example, in high-resolution spectroscopy of the Lyman-α forest in quasars, it may not be too important *which* sections of spectrum are observed, provided there is *sufficient* statistical information to identify velocity fields. Similarly, many statistical studies of galaxy counts etc. in direct images may place heavy demands on sample size, but not on sample continuity. In contrast, abundance studies requiring precise continuum placement (particularly in the presence of highly-blended lines, and with poor signal-to-noise ratio) will benefit if there are no gaps. For direct imaging there are obvious applications such as morphology of extended objects. It is true that these problems can be overcome by multiple exposures, but this reduces efficiency, and may prejudice the quality of the data if atmospheric conditions change. The upshot is that, whilst certain scientific programmes are reasonably transparent to butting gaps, others would benefit greatly by their reduction or elimination.

Applications of small mosaics (e.g. 2×2) include existing or planned echelle spectrographs, multi-object spectrographs and direct imaging. Tiling of large areas could open the way for major steps forward in, for example, large-scale direct studies of the structure of the universe, and medium-to-high resolution multi-object spectroscopy with large spectral coverage. This last was the subject of a major design study [Walker, 1989] which OSL undertook on behalf of the University of Columbia, New York. One solution explored was a mosaic-grating spectrograph (first order) with multi-fibre input, feeding a large mosaic of optimally-mixed CCDs in the focal plane of a large unfolded Schmidt camera. Mixed mosaics of solid state detectors is also an important aspect of the 'CWIC' (Complete Waveband Instrument Concept) being studied by Bingham *et al.* [1992].

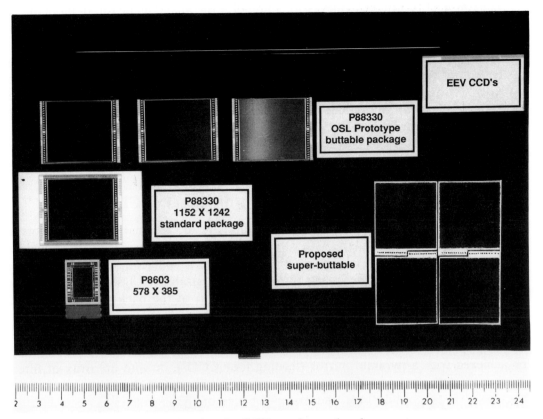

Figure 3. Mosaic CCD package development.

Recognising the advantages of mosaicing, OSL has collaborated with EEV on CCD package optimisation, with the emphasis on achieving the maximum degree of *4-way* buttability, in order to permit tiling of an indefinitely-large image area. This is in contrast to the work of several other groups who have concentrated on reducing the 'dead space' on two or three sides of a CCD package, restricting mosaicing to 2×2 or $2\times n$ respectively.

To achieve a significant step forward at minimal cost, a standard dice has been used – the P88330 1242×1152 format. The prototype package is shown in Fig. 3, adjacent to the standard package and, for comparison, the P8603 standard small-format device. In the prototype the dice bonding pads are brought out to pins *under* the active area, rather than alongside as in the standard package. In an indefinitely large mosaic this prototype achieves a butting gap along the short sides of 6.9 mm (cf. 28 mm in the standard package). The long sides achieve butting gaps of 1.8 mm (cf. the standard 7 mm). Note that the guard ring around the active area on the dice is retained and this is important to eliminate charge-injection from crystal lattice defects at the periphery of the dice.

The low-cost prototype achieves 75% filling of the image plane for an indefinitely large mosaic, which rises to 94% for applications such as multi-object spectroscopy where the individual spectra can be arranged to miss the wider gaps. If the package were simply scaled up to use the EEV 1152×2186 dice (a low-cost excercise), the filling factors would rise to 87% and 97% respectively. Larger gains are possible by *minor* changes to the mask-set, in order to re-route a simplified sub-set of the leadouts into one half of one side of the dice. The dice can then be jig-sawed together as shown in the cardboard model of the proposed 'super buttable', also in Fig. 3. Because the active area of the dice is unchanged, this is a low-risk solution, although the capital investment needed is significant due to the changes to the mask-set.

5.3. Optics to overcome butting gaps with mosaic CCDs

We have also explored optical methods to eliminate the residual butting gaps with the EEV CCDs, or larger gaps with other devices. Some of the ideas are shown in Fig. 4. Conceptually, the simplest method is to use a split coherent fibre bundle, which stacks sections of the image format onto separated CCDs. In practice this would not be used (except in the special application of the back-end of a photon counting system) due to the low efficiency.

Another generic method is to segment the focal plane of an instrument optically, and then re-image onto multiple CCDs with separate cameras. The segmenting may be achieved with a pyramid mirror (feeding four CCDs), or with an array of tilted mirrors, or with an array of wedge prisms. The last introduce chromatic aberration, but this is in the pupil plane (increasing the size of the re-imaging optics slightly) rather than the image plane. By locating the segmenting optics slightly off the image plane, the segmentation becomes defocussed.

A special case of the segmented focal plane is possible with a Baranne white-light pupil spectrograph. Baranne's scheme [Baranne, 1972], has an intermediate image of the spectrum which falls on or near to a large spherical mirror (the 'collector'). This in turn projects a white-light pupil (i.e. an image of the grating) onto the final camera optics which are then of minimum size. This avoids the usual problem of the large polychromatic beam envelope which cameras normally have to accept. The camera, working on finite conjugates, then re-images the spectrum onto the detector. The novel idea here proposed is to assemble the collector as a mosaic of smaller rectangular spherical mirrors. These would segment the first image of the spectrum. The mirrors would be individually tilted in order to feed separate (small) cameras and individual CCDs. The mirror segments would be optimally coated for their respective wavelength ranges, as would the optics of the separate cameras. Indeed, the cameras might have individually optimised optical designs, which might be completely different (e.g. dioptric for the red, catadioptric for the blue). The mosaic collector mirror could be manufactured (possibly by diamond-turning) in nickel-coated aluminium, since this would allow minimal bevelling cf. glassy materials.

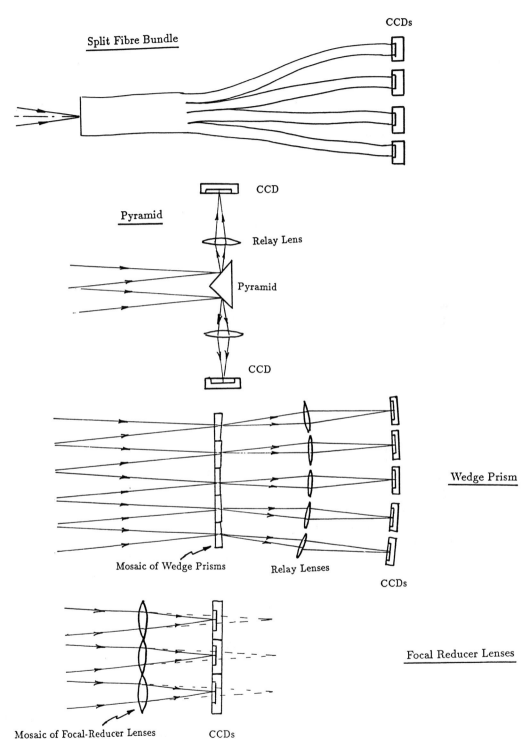

Figure 4. Some optical methods to overcome butting gaps with mosaic CCDs.

Another optical segmenting method we propose is to assemble in the image plane of a telescope or an instrument, a mosaic of rectangular lenses. The lens dimensions would be matched to the CCD *packages*, and would each de-magnify the image scale to fit the CCD *active areas*. This has the interesting feature that there is overlap between the fields recorded by adjacent CCDs, since the light from an object falling into the gap is shared between CCD active areas. A. Radley [1991] has undertaken a preliminary ray-tracing analysis of the method, applied to the f/8 focus of a telescope, and achieved geometric spot sizes of 0.25 arcsec projected to the scale of an 8-m telescope. This is clearly worthy of further study. In the case of a spectrograph, the lenses would be separately designed and anti-reflection coated for their respective sections of the spectrum.

6. Data analysis

The advent of large-format and mosaiced CCDs brings about severe problems of data handling. These are compounded when combined with a third dimension of data (e.g. time, polarisation etc.).

Under the UK 'Transputer Initiative' supported by the UK Department of Trade and Industry and the SERC, we have been exploring the potential for applying parallel processing techniques, particularly to the reduction of spectroscopic data (such as from echelle spectrographs). Mills has demonstrated [Mills, 1992] the ease with which existing FORTRAN and C code can be ported to Transputers. His first experiments involved data transfers achieved by setting up local RAM disks in each Transputer memory and then using e.g. FORTRAN I/O to access data in files on the RAM disks. This proved to be inefficient due to all the system overheads involved in file maintenance. This led to the implementation of 'Transputer Farms'. This concept involves using low-level message-passing routines directly.

With this new approach, Mills found that many data reduction algorithms (e.g. Monte Carlo simulation, cosmic-ray locator for CCD data, and arc-line identification) could achieve near-ideal speed gains. However, the relatively low throughput of the overall hardware/software message passing, prevented the effective use of the Transputer Farm for some algorithms in which the CPU-loading per processed data-value was low. An example was the median calculator. A Gaussian fitting routine gave intermediate speed gains (a gain of a factor approximately three for four transputers).

These results demonstrate the ease with which multiple processors can be directed towards the problem of handling the large data sets we may anticipate from the new generations of instruments e.g. for 8-m class telescopes.

7. Conclusion

From the pragmatic definition of mosaicing 'splitting the problem into manageable sections', it is apparent that the concept pervades much of astronomical instrumentation, from the optics of the telescope and instrumentation, through detectors, to the reduction of the data. The ESO Very Large Telescope is perhaps one of the most extreme cases, where four complete telescopes will be 'mosaiced' with the capability of feeding a combined focus. Currently, this is the most feasible solution for obtaining a 16-m effective aperture.

Numerous past advances in the performance of instruments would have been impossible without mosaicing, although the concept may not have been fully recognised as a coherent philosophy. Similarly, the advent of 8-m class telescopes is forcing the renewed exploration of mosaicing as a way to control costs and risk. It is hoped that this paper will inspire a fresh look at instrumental development from this perspective.

References

[Baranne, 1972] Baranne A., *Proc. ESO/CERN Conference on Auxiliary Instrumentation for Large Optical Telescopes, Eds Lausten S. & Reiz A., 1972.*

[Bingham *et al.*, 1990] Bingham R.G., Walker D.D., Diego F., *Proc. SPIE, Vol. 1236 (1990).*

[Bingham *et al.*, 1992] Bingham R.G., Walker D.D., Diego F., *Extending the Wavelength Coverage of Spectrographs. ESO Workshop on High Resolution Spectroscopy with the VLT, ESO, 1992, in press.*

[Boksenberg, 1971] Boksenberg A., *Astronomical Use Of Television-Type Sensors. Proc. Symp. at Princeton University, Princeton, N.J., NASA SP-256 (1971), p. 77.*

[Boksenberg & Burgess, 1973] Boksenberg A., Burgess D., *Astronomical observations with television-type sensors. Proc. Symp. at University of British Columbia, Vancouver, B.C., 1973, p. 21.*

[Boyle & Smith, 1970] Boyle W.S., Smith G.E., *Bell Syst. Tech. J., Vol. 49 (1970), p. 587.*

[Dierickx & Zigman, 1991] Dierickx P., Zigmann F., *The Messenger, ESO, No. 65 (1991), p. 66.*

[Epps, 1990] Epps H.W., *Proc. SPIE, Vol. 1235 (1990).*

[McLean, 1989] McLean I.S., *Electronic and Computer Aided Astronomy, Ellis Horwood, 1989, p. 83.*

[Mills, 1992] Mills D., *DTI/SERC Transputer Initiative Loan Report TR175, 1992.*

[Morgan, 1979] Morgan B.L., *Some Notes on the Application of CCDs in Astronomy., SERC SCOT paper SP/309/03, Nov. 1979.*

[NSF, 1989] *The NOAO 8-m Telescopes: III. Instrumentation. Proposal to the National Science Foundation, 1989, p. 11.*

[Nelson & Mast, 1990] Nelson J.E., Mast T.S., *Proc. SPIE, Vol. 1236 (1990).*

[Radley, 1991] Radley A., *An Investigation into Some Aspects of the Development of a Large Scale CCD Mosaic Imaging System for use in Space Applications. Report for MSc in Spacecraft Technology and Satellite Communications, UCL, 1991.*

[Rozelot & Leblanc, 1991] Rozelot J.P., Leblanc J.M., *Proc. SPIE, Vol. 1494 (1991).*

[Vogt & Penrod, 1987] Vogt S.S., Penrod G.D., *Proc. 9th Santa Cruz Summer Workshop 'Instrumentation for Ground-Based Optical Astronomy, Ed. Robinson L.B., 1987.*

[Walker, 1987] Walker D.D., *Proc. 9th Santa Cruz Summer Workshop 'Instrumentation for Ground-Based Optical Astronomy', Ed. Robinson L.B., 1987.*

[Walker, 1989] Walker D.D., *Preliminary Design Study of the Spectrograph for the Columbia University Wide Field Spectroscopic Telescope. Study commissioned by the Columbia University of New York, 1989.*

[Walker et al., 1985] Walker D.D., Sanford P., Lyons A., Fordham J., Bone D., *Advances in Electronics and Electron Physics, Vol. 64A (1985), p. 185.*

[Walker et al., 1990] Walker D.D., Bingham R.G., Diego F., *Proc. SPIE, Vol. 1235 (1990).*

9
Increasing the Efficiency and Range of Spectrographs

C.G. Wynne *

Abstract

Three aspects of improving the efficiency of general purpose spectrographs are discussed. First, new forms of camera, giving enhanced throughput at high apertures and field angles are described, together with their relevance to spectrograph layout. The second aspect deals with the properties and possible methods of use of immersed gratings in providing higher spectral resolutions and improved throughput, in a general purpose spectograph. The third describes the use of immersed gratings in tandem to give resolutions comparable to those of an echelle spectograph.

1. Introduction

General purpose spectographs are a major tool for astronomy but many of them around the world are shockingly inefficient. It is not uncommon for up to half the light passing the slit to be lost. In the present hard times I think efficiency matters. Considering the shortage of observing time, the shrinking funding for astronomy and the high cost of photons from faint objects, it would help all round to improve instrument throughput.

Over the last year or two, I have been working on various approaches to this. I have known for a long time about light loss in spectrographs; a number of those now in use are old optical designs of mine. But I did not realise until recently how much scope there is for extending the range of spectrographs in simple ways, which is one way of improving efficiency. So I am going to describe briefly three aspects of the work I have been doing, which turn out to interlock.

There are rather few people doing new thinking about these problems. Most noteworthy are the ESO Group, who, over the past few years, have published many very interesting and ingenious papers on spectrograph optics, mainly on lines quite different from mine. Perhaps the main difference is that I always have in mind that every air–glass surface, and still more every mirror, causes some loss of light; and these add up. So I have concentrated on simple solutions, which can also be more efficient.

*Institute of Astronomy, Madingley Road, Cambridge CB3 0HA, UK.

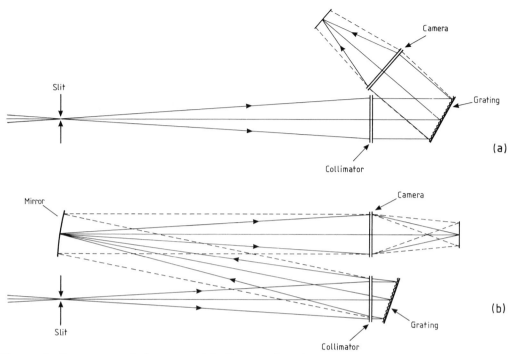

Figure 1. Spectrograph layouts: (a) Conventional spectrograph layout; (b) "White pupil" layout.

2. Cameras

The first aspect I am going to discuss is the camera. The major problem in the design of a conventional spectrograph (Fig. 1a) is the camera, which images the spectrum on the detector. The difficulty is that the aperture stop of the spectrograph camera lies on the grating, some distance in front of the camera, whereas all normal imaging systems (photographic lenses and the like) have an internal stop, and would not work otherwise. This requirement gets harder to satisfy as larger telescopes and multi-object spectroscopy call for faster cameras, covering wider fields of view – and I do not know of any existing spectrograph camera that works really satisfactorily under these conditions. Over the last 20 years when larger telescopes have become more common, various cameras have been designed and used – initially classical Schmidt cameras, or the sort of thing I designed for the AAT in 1972 (Fig. 2a)[Wynne, 1972], and related things for Kitt Peak and elsewhere. These all suffer from quite heavy vignetting, and rather poor imagery. More recently [Wynne, 1977] I found a major modification of the Schmidt type camera that has been used for all the larger spectrographs on the Isaac Newton Group of telescopes, La Palma (Fig. 2b). These work well for a single object, but would be hopelessly obstructive for multi-object work.

104

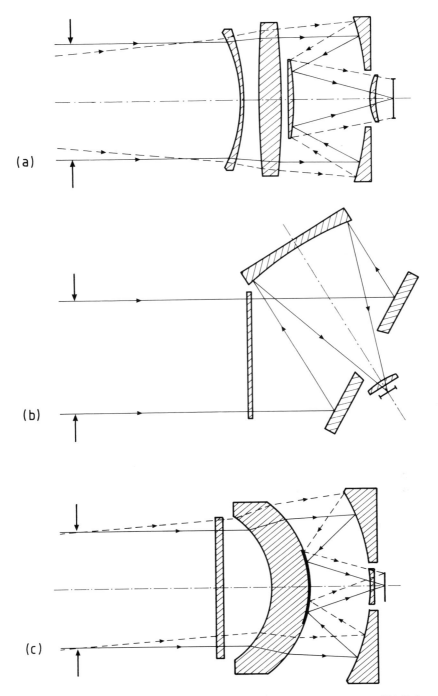

Figure 2. Section drawing of spectrograph cameras: (a) 25-cm f/1.7 in use on the Anglo-Australian Telescope [Wynne, 1972]; (b) 50-cm f/3.3 "Shorter than a Schmidt" design, in use on ISIS spectrograph on the 4.2 metre William Herschel Telescope [Wynne, 1977]; (c) 25-cm f/1.7 illustrating a new form of camera [Wynne, 1990].

105

It seems to have been assumed by the ESO group and others that an efficient camera with external stop was impossible, and in order to circumvent the problem they have adopted an early suggestion by Baranne (1965), which adds a further stage of imagery to the spectrograph. In this, called a "white pupil" system, the collimator, used in double pass, forms a large real intermediate image of the spectrum on a field mirror (or lens) adjacent to the slit (Fig. 1b). This intermediate image of the spectrum is then re-imaged by a relay lens on to the detector, with the field mirror forming an image of the grating (the stop) on to relay lens. This inevitably makes a larger and more complicated system than the classical spectrograph. And the use of the collimator in double-pass must give rise to some cross-talk between the white light from the slit and the diffracted light from the grating, to degrade the final image. This "parasitic light" problem has plagued white pupil designs from their beginning. It will obviously be more serious at high dispersions, when the illumination of the spectrum will be most highly attenuated, and more severe for multi-object work, with more sources to be back-reflected. Baranne [1988] cited various reasons for what he described as the "scientific failure" of the Casshawec spectrograph for the Canada-France-Hawaii Telescope. Among these was the fact that the use of multi-slit spectroscopy was not foreseen. Various steps can be taken to reduce stray light effects, but these inevitably lead to further complications.

All this arises from a failure to design a camera that works well with an external stop. I have been working on various different forms of camera and I have devised one that works better with a widely-spaced external stop than with an internal one, and it has other merits. It can be designed with a smaller central obstruction than earlier forms, it can cover quite wide field angles, without vignetting, at high relative apertures – and it has greater stability of aberration correction with wavelength than anything else I know. And all this involves no greater complexity than existing systems. Fig. 2c shows an example, at f/1.7, data for which were published last year [Wynne, 1990] – and Fig. 3 shows spot diagrams for a further example, at f/1.2 [Wynne, 1991a]. My colleague I. Escudero-Sanz, has shown how to get similar results in an even simpler system [Escudero-Sanz, 1992]. These publications all give numerical data and some explanation of how and why the system works. And the existence of these designs should relieve the problems of light loss in spectrograph cameras. We shall come back to these cameras in a different context later on.

3.   Gratings

The second aspect of efficiency concerns gratings and their interaction with cameras.

The classical spectrograph works over a wide range of wavelengths and dispersions using interchangeable gratings with variable obliquities. At low dispersions, with coarser gratings and small obliquities, the diffracted beam is only slightly longer than the incident one (Fig. 4a); with finer gratings at larger obliquities the diffracted beam expands, first to fill and then to overfill the grating (Fig. 4b) with direct loss of light, increasing with obliquity. It is this effect that limits the use of the conventional

106

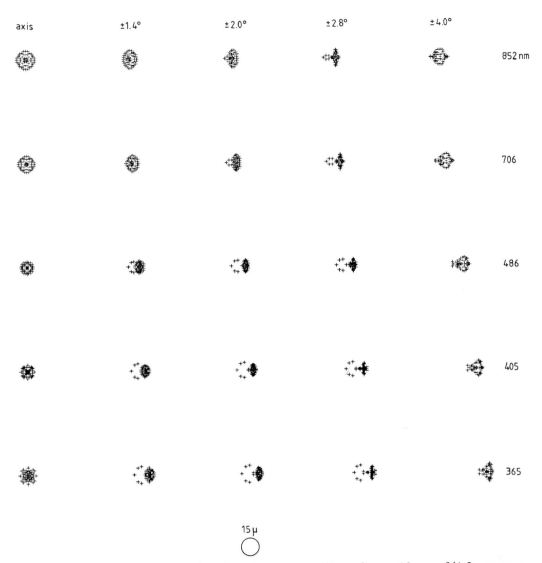

Figure 3. Spot diagrams showing image quality of an 18-cm f/1.2 camera [Wynne, 1991a].

spectrograph at higher spectral resolutions. So I have been looking at different ways of using gratings, and it turns out that immersing a grating to an appropriate prism makes it possible to achieve a spectral resolution about double that obtainable in air.

The history of immersed gratings is strange. In 1954 Hulthén and Neuhaus, in a letter to *Nature*, described two immersion experiments giving discordant results which they did not explain – and no-one seems to have pursued the matter until

107

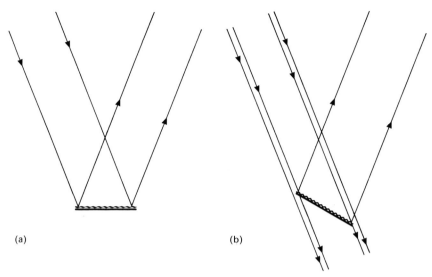

Figure 4. Light of wavelength 800 nm, arranged to give 40° between incident and diffracted beams (a) on a grating having 75 grooves/mm; (b) on a grating having 1200 grooves/mm.

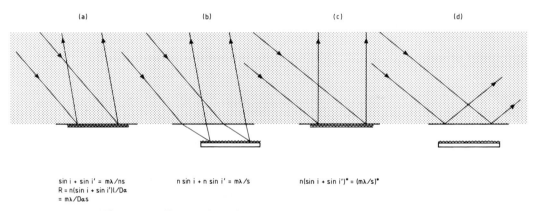

Figure 5. Illustrating the theory of immersed gratings.

1987, when Dekker announced as a general principle that immersion increased resolution n times (n being the refraction index). This is certainly not generally true, but there are rather peculiar conditions when it is. There has been some confusion about this, so I shall explain briefly what happens.

Consider diffraction at an immersed grating (Fig. 5a) which follows the standard grating equation

$$\sin i + \sin i' = m\lambda/ns \qquad (1)$$

where i and i' are angles at the grating, λ the wavelength in air, n the index, s the

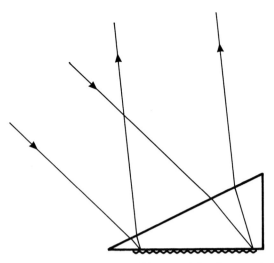

Figure 6. Light of wavelength 700 nm, diffracted in first order by a grating of 2400 grooves/nm immersed to a prism.

groove spacing, m the order. This leads to the well-known expression giving the resolution R when the grating is used on a telescope

$$R = \lambda/\delta\lambda = n(\sin i + \sin i')l/D\alpha = m\lambda l/D\alpha s \qquad (2)$$

where α is the seeing angle, D the entrance pupil diameter, and l is the length of the beam intersection with the grating plane, in the dispersion azimuth. If we now separate the grating from the substrate, so that light reaches it through a plane parallel air space (Fig. 5b), the angles of incidence and the wavelength are increased by a factor n, giving

$$n(\sin i + \sin i') = m\lambda/s \qquad (3)$$

which is identical with the equation for the immersed grating. R is unaltered.

This argument breaks down if i or i' is greater than the critical angle of the medium, in which case light cannot reach the detached grating, but is totally reflected (Fig. 5c). The immersed grating is then working under conditions unattainable in air, which can yield higher resolutions. For example, at a given wavelength a finer grating can become usable. These effects arise of course from the reduced wavelength on immersion. In addition to immersion there are two smaller side effects from the prism that increase the resolution further, giving overall an approximate doubling of resolution.

Fig. 6 shows the form of a typical immersed system and Table 1 shows a comparison at various wavelengths of a 1200 grooves/mm grating in air, and an immersed grating of 2400 grooves/mm. The increase in R comes from the finer grating, the use of which is made possible by immersion. The column 0 gives, where

109

Table 1. Grating Parameter Comparison

λ (nm)	Grating 1200 g/mm, m = 1, in air			Grating 1200 g/mm, m = 2, or 2400 g/mm, m = 1, immersed		
	l (cm)	0	R	l (cm)	0	R
340	18.3	-	3727	18.3	-	7446
500	19.6	-	5876	20.1	-	12062
600	20.9	1.5%	7528	19.4	-	14000
700	22.4	8.0%	9404	20.6	-	17295
800	24.3	15.3%	11674	22.1	6.7%	21197
900	26.9	23.4%	14526	24.1	14.3%	25974

$D = 4m$; grating 15.4×20.6 cm; $\alpha = 1$ arcsec.

relevant, the length of the overfill of the grating, as a percentage of the beam length in the dispersion azimuth.

The effect could be applied at various levels. Fig. 6 could be interchangeable on an existing spectrograph (ISIS on the William Herschel Telescope or IDS on the Isaac Newton) to give double the resolution otherwise available, over a range of wavelengths, at a trivial cost for such a gain, and this could subsequently be applied more widely to include coarser gratings; the output beam diameters at high and low dispersions could be equalised with a gain of throughput overall. A fuller account has been given in [Wynne, 1991b].

4.   Tandem gratings

The work thus far has shown that simple and relatively inexpensive changes to a classical spectrograph can bring very substantial increases of both throughput and range of dispersion. The third aspect I have been looking at, which hangs on the first two, relates to still higher resolutions.

For higher resolutions than conventional spectrographs can give, large echelle spectrographs are at present used. These have various disadvantages arising mainly from the need for cross-dispersion to separate their output of successive overlapping orders of short free spectral range. They are very expensive, large and heavy, requiring location at coudé or Nasmyth focus with consequent light loss; usable slit-length is limited and the multi-object spectroscopy impossible.

Their advantage is in giving higher resolutions than have been otherwise available, and this derives from two features. First, their very coarse rulings mean that they can be ruled in larger sizes – up to 40.8 cm – double what is generally available for finer gratings. Secondly, their use in near-Littrow configuration makes larger values of $(\sin i + \sin i')$ possible, by a factor of about 1.7 (see equation 2). Thus the

110

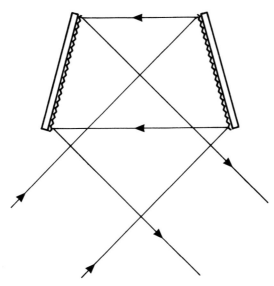

Figure 7. Two air-spaced gratings in tandem (after [Bingham, 1983]).

inherent maximum R of an echelle is about 3.4 times that of the classical Cassegrain spectroscope. This factor of about 3.4 for the echelle advantage seems at first sight rather small; this is perhaps due to the fact that figures for the Cassegrain spectrograph are generally given for a one arcsec slit, for an echelle for a half arcsec. In both types, equally, R can be increased by a reduction of slit-width up to the diffraction limit.

The use of immersed gratings would then reduce the echelle advantage to a factor around 1.7. To equal the echelle performance, the classical form needs a further increase in dispersion.

An obvious way to double dispersion is to duplicate the disperser. This was done by Huggins & Miller [1863] by using two prisms in place of one in an astronomical spectrograph. Tull [1972] used a reflection grating in double-pass. Baranne [1987] used two transmission gratings in tandem. Bingham [1983] suggested the use of two reflection gratings, but did not develop it into a spectrograph design (Fig. 7).

With two air-spaced fine-ruled gratings, an arrangement that can be tuned to cover a range of wavelength does not seem possible, but this can be achieved using immersed gratings in tandem. Fig. 8 shows an example using two gratings of 1200 grooves/mm in second order (or 2400 in first) immersed to 24° prisms. This system can be tuned, over a range of wavelength, by tilting the gratings about the prism vertices. Figs 8a, 8b and 8c show the system tuned for central wavelengths of 600, 700 and 800 nm. On a 4-m telescope, with a 15.4-cm collimated beam and a 1 arcsec slit, this would give R values of 30 000 at 600 nm, 38 000 at 700 nm and 52 000 at 800 nm. For a 63.5° echelle under the same conditions the R (independent of λ)

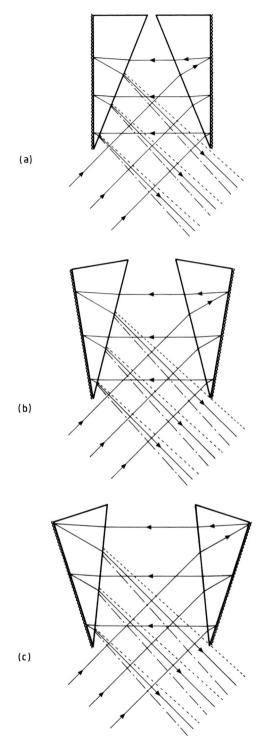

Figure 8. Configurations of two immersed gratings in tandem, tuned for use (a) at 600 nm, (b) at 700 nm, (c) at 800 nm.

would be 37645. The apertures of input and output beams are equal, at all settings, eliminating a serious cause of inefficiency in classical spectrographs.

This use of two immersed gratings in tandem produces an exit pupil close behind the second grating, requiring a camera working with a front stop distance of about 2.7 times the beam diameter. This would be a serious obstacle to the use of these systems, but that recent work on new camera designs discussed above provides a solution.

These tandem systems, giving about four times the resolutions of classical spectrographs, are more complicated to the extent of a second prism, involving additional light loss of some 30 per cent. But they share the advantages of the classical systems of a size suitable for location at Cassegrain focus and of providing direct access to a wide unencumbered free spectral range, usable over a long slit, and for multi-object spectroscopy. They appear to constitute a useful alternative to echelle spectrographs.

References

[Baranne, 1965] Baranne A., *Compt. Rend. Acad. Sci. Paris, Vol. 260, (1965), p. 2383.*

[Baranne, 1987] Baranne A., in *Instrumentation for Ground-Based Astronomy.* ed. Robinson, L.R., (1987), Springer-Verlag, New York, p. 296.

[Baranne, 1988] Baranne A., in *ESO Conference on Very Large Telescopes and their Instrumentation.* ed. Ulrich, M.-H., (1988), ESO, p. 1195.

[Bingham, 1983] Bingham R.G., *SPIE, Vol. 448, (1983), p. 416.*

[Dekker, 1987] Dekker H., in *Instrumentation for Ground-Based Astronomy.* ed. Robinson, L.R., (1987), Springer-Verlag, New York, p. 183.

[Escudero-Sanz, 1992] Escudero-Sanz I., *J. Mod. Optics, Vol. 39, (1992), p. 711.*

[Huggins & Miller, 1863] Huggins W., Miller W.A., *Proc. Roy. Soc., Vol. 12, (1863), p. 444.*

[Hulthén & Neuhaus, 1954] Hulthén E., Neuhaus H., *Nature, Vol. 173, (1954), p. 442.*

[Tull, 1972] Tull R.G., in *ESO/CERN Conference on Auxiliary Instruments for Large Telescopes, eds Lausten S. & Reiz A., (1972), Geneva, p. 259.*

[Wynne, 1972] Wynne C.G., *Mon. Not. R. astr. Soc., Vol. 157, (1972), p. 403.*

[Wynne, 1977] Wynne C.G., *Mon. Not. R. astr. Soc., Vol. 180, (1977), p. 485.*

[Wynne, 1990] Wynne C.G., *Mon. Not. R. astr. Soc., Vol. 247, (1990), p. 173.*

[Wynne, 1991a] Wynne C.G., *Mon. Not. R. astr. Soc., Vol. 252, (1991), p. 171.*

[Wynne, 1991b] Wynne C.G., *Mon. Not. R. astr. Soc., Vol. 250, (1991), p. 796.*

10

The Problem of Achromatizing Astronomical Optics for the Near and Mid Infrared

I. Escudero-Sanz *

Abstract

The designer of optical instruments for astronomy must bear in mind two ultimate requirements: high efficiency and achromatism over a wide spectral range. The problem of achromatization becomes more acute in the near and mid infrared because this spectral interval is wider and the number of available materials with good transmission is smaller than in the visible. The present work is concerned with this problem and some solutions to it are presented.

1.   Introduction

The optical instruments which are currently in demand for astronomy in the near- and mid-infrared can be classified in three groups: spectrographs, low-resolution imaging and high-resolution imaging cameras. The appropriate plate scales go from 1 down to 0.1 arcsec per pixel. For the two types of detectors which are currently used (InSb arrays and HgCdTe arrays, with typical pixel sizes varying between 30 and 60 μm approximately) the final focal ratios required go from the very fast (f/1.3 or faster) to the very slow (f/30 or slower).

In general, the aberrations which must be corrected in an infrared instrument are the same as those of an optical instrument: the Seidel spherical aberration, coma, astigmatism and field curvature, together with the high orders. The correction of the chromatic aberrations is also important: any instrument must be free of chromatic difference of focus (CI) unless the detector is moved back and forth to refocus for each wavelength; the chromatic difference of magnification (CII) is usually irrelevant in spectrograph cameras but should be corrected in imaging systems. Often, the direct consequence of correcting CI and CII is that the high-order chromatic aberrations set the limits on the wavelength range and numerical aperature at which an optical system can work. Sometimes it is impossible to

*Institute of Astronomy, Madingley Road, Cambridge CB3 0HA, UK.

114

achromatize the optics by using materials with different dispersions due to either the mismatch of refractive indices and dispersions or the large secondary spectrum introduced. Any solution to a particular design problem analyzed should have the minimum possible number of optical surfaces so as to keep a high efficiency. Some additional complications appear in the infrared: not only does the baffling have to be different but also the optics have to be cooled. Compactness and simplicity are therefore fairly desirable characteristics if the cost of the cryogenics is to be reduced.

Amongst the examples included in this paper there are three spectrograph cameras (one of which could be part of an imaging system) in which the solution to the problem of achromatization maintaining a good efficiency is to combine mirrors and lenses. The other two systems presented are intended for imaging instruments and show what can be done with optical glasses which have good transmission in the near-infrared.

2. ZnSe-based cameras for spectrographs

Some infrared detectors are able to register photons with wavelengths between 1 and 5 μm. Fortunately, the designer can find optical materials with good transmission in the complete spectral range which open the door to the possibility of using both refractive and reflective components in a design, a practice common enough in astronomical instrumentation. An example is ZnSe, a material with a high refractive index which can be very useful, not only in catadioptric systems like the ones used in the visible (such as the Schmidt or Maksutov cameras), but also in some other more simple and compact systems. The two designs described in this section are corrected for longitudinal colour (CI) for a wavelength interval of 4 microns (from 1 to 5). The correction is based on the use of refractive components which are self-achromatic; this simply means that they do not introduce any CI individually. Both systems could be implemented in spectrographs. Their design parameters are presented in a scale such as to accept a collimated beam of 15 cm diameter.

The first system is an f/1.3 camera which has only two optical components: a mirror and a lens near to focus. The lens acts as a field flattener for the mirror. The correction of the Seidel aberrations is achieved with the help of two aspheric surfaces: the mirror itself and the surface of the lens closer to the mirror. (It is the achromatized version of an infrared camera by [Attad-Ettedgui & Mountain, 1988].) The data and a scaled drawing are given in Table 1 and Fig. 1(a). For the scale used, images of stellar objects smaller than a 50-μm diameter circle are obtainable for a full field of view of 4.6 deg, (or images smaller than 35 μm for a full field of view of less than 3.2 deg: an equivalent linear field of 12 mm) as shown in the spot diagrams in Fig. 1(b).

Table 2 and Figs 2(a), 2(b)1 and 2(b)2 contain the data, scale drawing and spot diagrams of an f/1.7 camera whose optical components are also mirrors and ZnSe lenses: a concave mirror and a self-achromatic meniscus with an aspheric

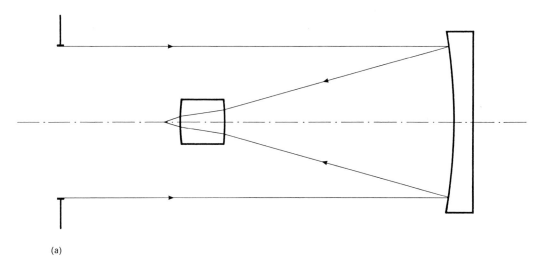

(a)

Figure 1. (a) Scaled drawing of the f/1.3 infrared spectrograph camera.

Table 1. f/1.3 infrared spectrograph camera with a ZnSe lens.

Curvature radius (cm)	Separation (cm)	Material (cm)	Clear diameter (cm)
Plane (stop)			15.0
	−38.800	air	
53.064*			18.0
	22.475	air	
26.484†			4.4
	4.307	ZnSe	
−18.712			4.4
	1.393	air	
Plane (focus)			

The equivalent focal length is 19.5 cm. Surfaces with positive curvature are concave to the stop.
* Aspheric surface, departure from sphere: $z = 1.1008 \times 10^{-6}h^{-4} - 1.164 \times 10^{-9}h^{-6}$.
† Aspheric surface, departure from sphere: $z = 1.8635 \times 10^{-3}h^{-4} - 1.094 \times 10^{-4}h^{-6}$.

surface some distance in front, plus a secondary mirror which coincides with the back surface of the meniscus and a self-achromatic lens near to focus. This system is not as simple as the previous one but is very compact and has an accessible focus. For the spectral region between 1 and 5 μm it can provide images of star objects

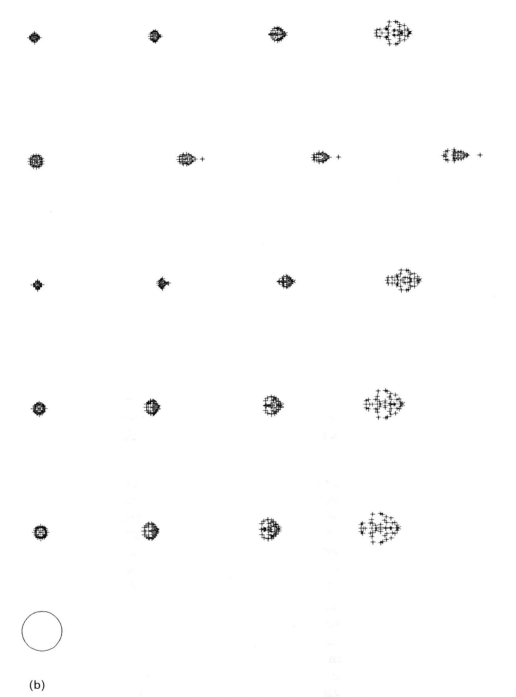

(b)

Figure 1. (b) Spot diagrams of the f/1.3 infrared camera. The spots correspond to 0, 1.1, 1.6 and 2.3 deg field angles from right to left and the wavelengths 2.2, 1.0, 1.8, 3.8 and 5.0 μm from top to bottom. The circle has a diameter of 50 μm. The images are smaller than a 35 μm diameter circle within a full field of view of 3.2 deg.

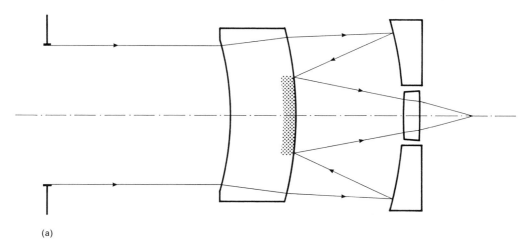

(a)

Figure 2. (a) Scaled diagram of the f/1.7 infrared spectrograph camera. (b)1 Spot diagrams of the f/1.7 infrared camera at a back focal distance of 5.5435 cm. The spots correspond to 0, 1.4 and 2.1 deg field angles from right to left and the wavelengths 2.2, 1.0, 1.8, 3.8 and 5.0 μm from top to bottom. The circle has a diameter of 35 μm. (b)2 Spot diagrams of the f/1.7 infrared camera at a back focal distance of 5.5450 cm. The spots correspond to 0, 1.0 and 1.4 deg field angles from right to left the wavelengths 2.2, 1.0, 1.8, 3.8 and 5.0 μm from top to bottom. The circle has a diameter of 35 μm.

smaller than a 50-μm diameter circle for a field of view of 4.2 deg at a back focal distance of 5.5435 cm (or, alternatively, images smaller than 35 μm within a 2.8 deg field at a back focal distance of 5.5450 cm; this is a 13-mm linear field). The total vignetting caused by the central obscuration of the secondary is less than 40% of the area (assuming a working field of 4.2 deg, as in Table 2).

A remark about the feasibility of these two cameras must be made at this point. Some of the aspheric surfaces could be difficult to manufacture with the traditional polishing procedures; however, since the tolerances of surface roughness are less tight for longer wavelengths and since materials like ZnSe can be diamond-turned, such a difficulty should be easily overcome in the infrared.

3. A very fast camera

Some optical glasses transmit near-infrared light. The catalogue of Schott includes 14 optical glasses with transmission better than 96% between 1 and 2.3 μm for a thickness of 5 mm. These materials could be used in instruments with infrared detectors which work only in the near infrared. The system described below is an example of an f/1.2 camera which has two lenses of the optical glass LaSF35 from Schott. It is a modified Schmidt–Maksutov camera that can work at higher numerical apertures and bigger field-angles than the classical Maksutov or Schmidt types. This

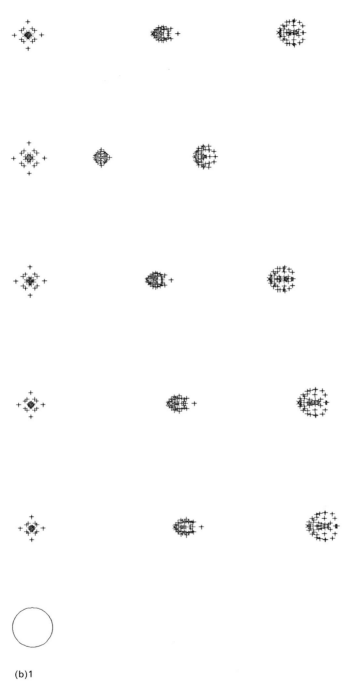

(b)1

Figure 2. *Continued*.

119

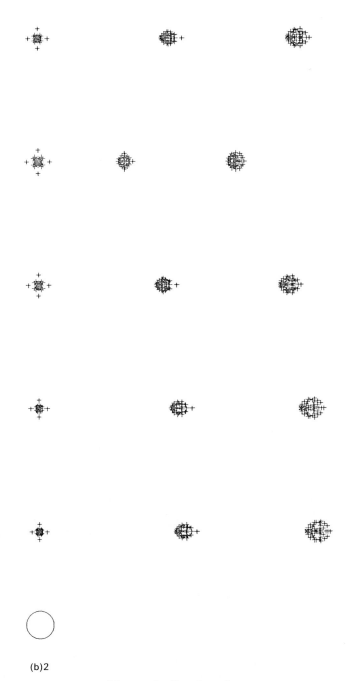

(b)2

Figure 2. *Continued.*

120

Table 2. f/1.7 infrared spectrograph camera with ZnSe lenses

Curvature radius (cm)	Separation (cm)	Material (cm)	Clear diameter (cm)
Plane (stop)			15.0
	20.000	air	
−30.357*			16.4
	7.040	ZnSe	
−37.931			19.0
	11.293	air	
−39.171			20.6
	−11.293	air	
−37.931			10.3
	11.293	air	
40.476			5.6
	1.821	ZnSe	
−100.990			5.2
	5.545	air	
Plane (focus)			

The equivalent focal length is 25.68 cm. Surfaces with negative curvatures are concave to the stop.
*Aspheric surface, departure from sphere: $z = -2.1185 \times 10^{-6} h^{-4} - 2.20 \times 10^{-9} h^{-6}$.

system consists of three optical components: a concave mirror, a meniscus with an aspheric surface which corrects the mirror aberrations, and a self-achromatic field-flattener lens near to focus. The data, scaled-drawing and spot diagrams can be found in Table 3 and Figs 3(a) and 3(b). Again a stop of 15-cm diameter has been assumed. It provides images of stellar objects smaller than a 50-μm diameter circle within a full field of view of 10 deg (or better than 40 μm within a full field of view of 7 deg; this is a linear field of 11 mm). It should be noted than both CI and CII are corrected and therefore the design could be used for imaging purposes as well as for spectroscopy.

4.   Using optical glasses with different dispersions

The main problem in the design of refractive optics for the infrared is the paucity of materials with good transmission. Hence it is difficult to find materials with refractive indices and dispersions appropriate to fit in achromatized sets. It has been mentioned above that some optical glasses have good transmission in the near infrared. These may be useful in the instruments intended for the J, H and K atmospheric windows, either together with the materials traditionally used in this region – such as CaF_2,

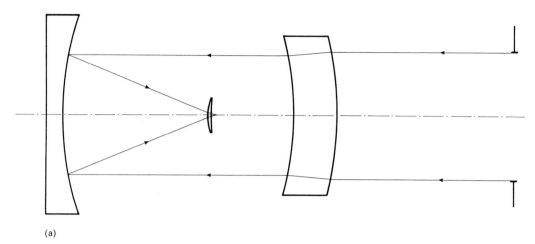

(a)

Figure 3. (a) Scaled drawing of the f/1.2 camera for the near infrared.

BaF_2, MgO, LiF, etc. [Wolfe, 1978] – or in combinations of themselves. The optical glasses would have advantages in the processes of manufacturing, aligning and testing because these tasks can be performed with visible light.

Two examples of the author's current work on this matter are included here. They are cemented triplets which have been designed for small imaging instruments and would accept stops of 2 cm diameter. The Schott optical glasses $BaSF_2$ and SFL56 have been used. In both systems CI and CII are corrected within full fields of view corresponding to an image surface of 10 mm diameter. The first example is an f/13.5 camera which would provide images of stellar objects smaller than 50 μm (see Table 4 and Fig. 4). The second system has an aspheric surface which contributes to the correction of the monochromatic aberrations; in this way the triplet can be slightly faster (f/12) while providing improved imagery (images of stellar objects smaller than 35 μm). The data and spot diagrams for this triplet are given in Table 5 and Fig. 5 respectively.

5.   Conclusions

The problems related to achromatization faced by the designer of optics for the near- and mid-infrared increase since wider spectral ranges must be covered. A further difficulty arises because of the paucity of materials with good transmission. However, there are means to achromatize systems which can be successfully implemented.

Sometimes the solution can be a catadioptric system. Three examples of this have been described which include lenses of ZnSe and the Schott optical glass LaSF35. When only refractive components are used the problem with achromatization is to find materials with the appropriate refractive indices and dispersions. In the near-infrared some optical glasses could be added to the list of materials with good transmission. This would facilitate the search for a good combination. Two

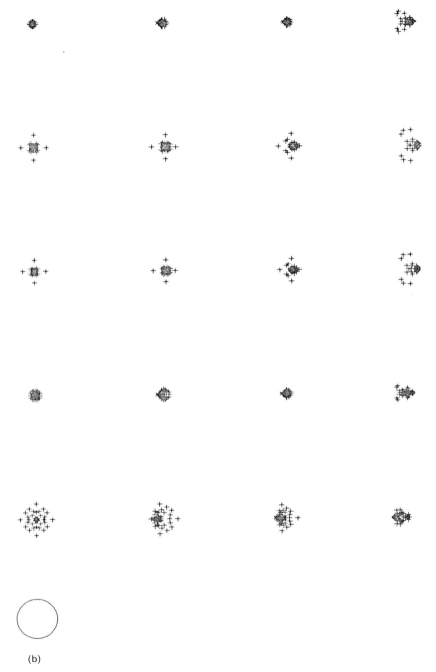

(b)

Figure 3. (b)Spot diagrams of the f/1.2 camera for the near infrared of Table 3. The spots correspond to 0, 2.4, 3.5 and 5.0 deg field angles from right to left and the wavelengths 1.5296, 1.0140, 1.0600, 1.9701 and 2.33540 μm from top to bottom. The circle has a diameter of 50 μm. The images are smaller than a 40 μm diameter circle within a full field of view of 7 deg.

123

Table 3. f/1.2 camera for the near infrared with lenses of optical glass LaSF35.

Curvature radius (cm)	Separation (cm)	Material (cm)	Clear diameter (cm)
Plane (stop)			15.0
	−20.757	air	
−40.287*			18.8
	−5.126	LaSF35	
−37.333			18.2
	−27.291	air	
35.948			23.6
	17.313	air	
6.598			4.2
	0.388	LaSF35	
27.580			4.2
	0.492	air	
Place (focus)			

The equivalent focal length is 17.51 cm. Surfaces with positive curvature are concave to the stop.
*Aspheric surface, departure from sphere: $z = 4.2736 \times 10^{-6}h^{-4} + 5.696 \times 10^{-9}h^{-6}$.

Table 4. f/13.5 cemented triplet for the near infrared with the optical glasses BaSF2 and SFL56.

Curvature radius (cm)	Separation (cm)	Material (cm)	Clear diameter (cm)
26.078 (stop)			2.00
	0.772	BaSF2	
−8.056			1.99
	0.340	SFL56	
8.218			2.00
	0.793	BaSF2	
−16.372			2.01
	26.630	air	
Plane (focus)			

The equivalent focal length is 26.58 cm. Surfaces with negative curvature are concave to the stop.

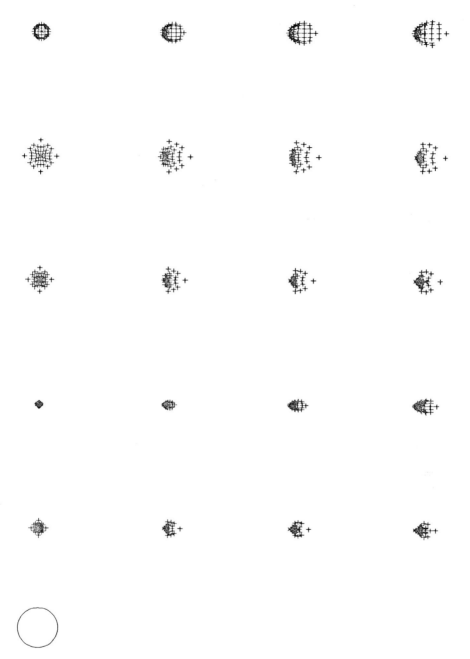

Figure 4. Spot diagrams of the f/13.5 triplet for the near-infrared made of the Schott optical glasses BaSF2 and SFL56. The spots correspond to 0, 0.59, 0.84 and 1.1 deg field angles from right to left and the wavelengths 1.5296, 1.0140, 1.0600, 1.9701 and 2.33540 μm from top to bottom. The circle has a diameter of 50 μm.

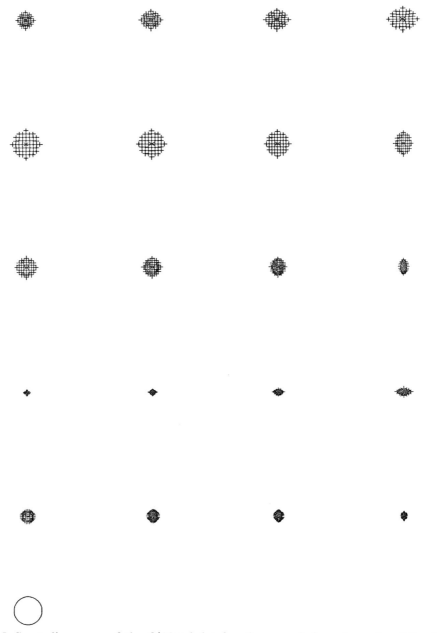

Figure 5. Spot diagrams of the f/12 triplet for the near-infrared made of the Schott optical glasses BaSF2 and SFL56. The spots correspond to 0, 0.56, 0.84 and 1.2 deg field angles from right to left and the wavelengths 1.5296, 1.0140, 1.0600, 1.9701 and 2.33540 μm from top to bottom. The circle has a diameter of 35 μm.

126

Table 5. f/12 cemented triplet for the near infrared with the optical glasses and BaSF2 SFL56.

Curvature radius (cm)	Separation (cm)	Material (cm)	Clear diameter (cm)
13.377 (stop)			2.00
	0.714	BaSF2	
−7.133			1.98
	0.179	SFL56	
7.164			1.97
	0.536	BaSF2	
−26.860*			1.97
	23.290	air	
Plane (focus)			

The equivalent focal length is 24 cm. Surfaces with negative curvature are concave to the stop.
*Aspheric surface, departure from sphere: $z = -8.8464 \times 10^{-5}h^{-4}$.

cemented triplets made with the Schott optical glasses $BaSF_2$ and SFL56 have been presented.

References

[Attad-Ettedgui & Mountain, 1988] Attad-Ettedgui E., Mountain, C.M., *Infrared Systems, Design and Testing. SPIE, Vol. 916 (1988), p. 27.*

[Wolfe, 1978] Wolfe W.L., *Handbook of Military Infrared Technology. Office of the Naval Research Department of the Navy: Washington, D.C.*

11
The Next Generation of Near Infra-Red Imaging Spectrometers

*A.K. Forrest * and M. Wells †*

Abstract

The advantages and disadvantages of various possibilities for imaging spectrometers on new-generation large telescopes are discussed, and the favourable prospects for the Michelson approach are detailed. A particular Michelson-spectrometer proposal for an 8-m telescope is outlined.

1.   Need for a new generation

For the last twenty or thirty years it has been possible to scale existing astronomical instruments as telescopes got larger. For example coudé grating spectrometers have grown in size with the telescopes. The latest generation of telescopes has, however, made this scaling very difficult. At the same time area detectors have advanced from being unusable because of poor efficiency, to providing the best obtainable quantum efficiency with spatial resolutions of 1000×2000.

The two advances of telescope size and array detectors, allowing spectrometer operation on large fields of view, have produced a need for instruments with enormous luminosity. It is no longer possible to scale existing instruments. The wavelength range of interest has extended from the visible to the infra-red, and all the capabilities of visible instrumentation are now expected in the infra-red. Infra-red instruments generally need to be cooled. This requires that the instrumentation is made small; it would not be possible to cool a conventional coudé spectrograph.

The laws of scaling are also unfavourable. The mass of an instrument goes up with its linear size cubed, but its strength goes up only as length squared. The required stability of the instrumentation stays the same for any size telescope. This will result, if no new approach is taken, in hugely massive and expensive instruments.

---

*Centre for Robotics and Automated Systems, ICSTM, Exhibition Road, London SW7 2BX, UK.
†Royal Observatory, Blackford Hill, Edinburgh EH9 3HJ, UK.

The spatial resolution required from new instruments is much greater than in the past. This is not just because of the availability of good area detectors. It has become obvious that very high spatial resolution on the sky is of great benefit. Very good imaging allows observation of stars in other galaxies and the reduced effective background allows the observation of much fainter objects. This is in contrast to the previous view, especially of infra-red telescopes, as flux buckets. Active and adaptive optics are pushing the technology further in the direction of high spatial resolution. It is fortunate that this is the case. Full benefit can be made of the area detectors which if given an arcsec on the sky per pixel would cover more of the sky than is usually useful, and might even start to get outside the angle of optimum performance of the telescope. These changes have put yet another constraint on new instruments, that is they must have very good imaging qualities.

2.   Methods for obtaining optical efficiency

For the purposes of this paper we shall use a rather unusual definition of efficiency, namely spectral resolution $\mathscr{R}_\lambda$, times spatial resolution $\mathscr{R}_x$, times luminosity L divided by the size S of the instrument.

$$\text{Efficiency} = \frac{\mathscr{R}_\lambda \mathscr{R}_x L}{S}$$

The possibilities are Grating, Fabry–Perot (F–P), Michelson or some more unusual and specialised possibilities such as heterodyne instruments.

The grating, F–P and Michelson at the deepest level are surprisingly similar. The Michelson is very similar to an F–P with finesse of two. The F–P has multiple beam interference and multiple orders, the same as a grating. The basic performance of the different systems is ultimately dependent on a few simple characteristics, for example the resolution is dependent on the maximum path difference between interfering beams. Similarly the way to get minimum size of optical components is to use them on-axis. On these very basic characteristics therefore the F–P and the Michelson give the best chance of providing a small high efficiency instrument.

The aim in our particular case is to produce a resolution $\mathscr{R}_\lambda = 2000$ instrument capable of working up to a wavelength of $5\,\mu$m. This implies a Michelson with maximum path difference of 10 mm, or a pair of F–Ps in series. A pair of F–Ps is needed because it is not possible to obtain a finesse of much better than forty for a high luminosity (and hence large) instrument.

3.   Fabry–Perot or Michelson

Rather than report a comprehensive comparison of the two instruments we will just state why we ended up choosing the Michelson.

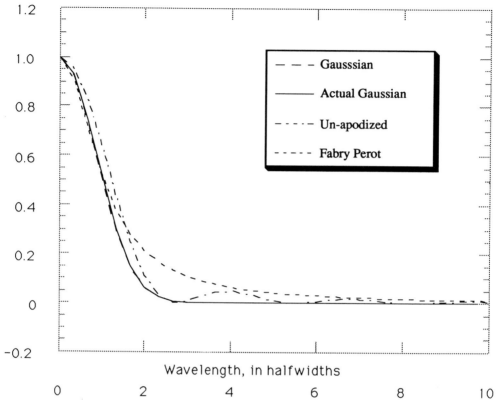

Figure 1. Instrumental profiles; the instrumental profile of the approximate Gaussian is indistinguishable from a theoretical Gaussian with this scaling.

3.1.   Instrumental profile

The Michelson has a very good instrumental profile (see Figs. 1 and 2). The wings of an un-apodized response are well under the broad wings of the F–P profile. By using apodizing (weighting samples in the interferogram differently), and using a slightly larger path difference than simple calculations suggest, it is possible to get, for example, a Gaussian instrumental profile with very low wings. In fact the spurious responses from this source can be made negligible compared to other effects such as ghost interferograms from air/glass interfaces, inaccurate sample spacing, etc. The case of apodization in Figs. 1 to 3 is to use a Gaussian function truncated at twice the nominal path difference required for this resolution. The response of the system to out-of-profile signals is three orders of magnitude (in power) better than the non-apodized case.

There is no penalty to be paid in signal-to-noise terms if the apodization is accomplished by observing the interferogram samples for different times rather than observing each point in the interferogram for the same time and then adjusting the

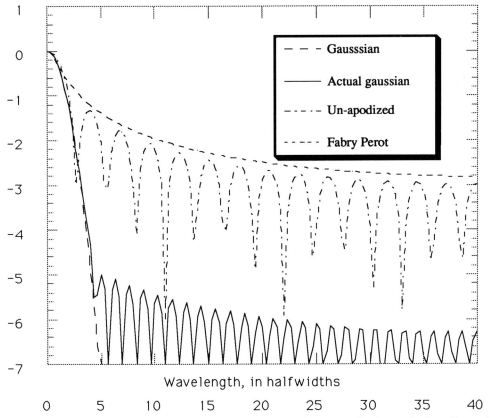

Figure 2. Logarithmic instrumental profiles. The superiority of the Gaussian apodized profile is very obvious in this plot. At the cost of scanning twice as far as normal the response off the line is well below effects due to stray light, ghost interferograms, etc.

results. The only price to be paid is an instrument with slightly larger path difference than expected. The equivalent pair of F–Ps arranged in a vernier system has quite large (5%) spurious response at some wavelengths and offsets and never has very high rejection of wavelengths outside the nominal pass band.

The instrumental profile of the Michelson is variable even during an observation. The apodizations in Figs. 1 to 3 represent only a couple of a very large number of possibilities, the best being dependent on the particular observation. For example although the demonstrated apodization has good rejection of out-of-band signals there are very deep nulls in the out-of-band response. These can be removed by replacing the hard truncation of the Gaussian apodization with a softer function. Whether to use any of these techniques is the decision of the observer, not the instrument builder. In contrast the instrumental profile of the F–P is fixed.

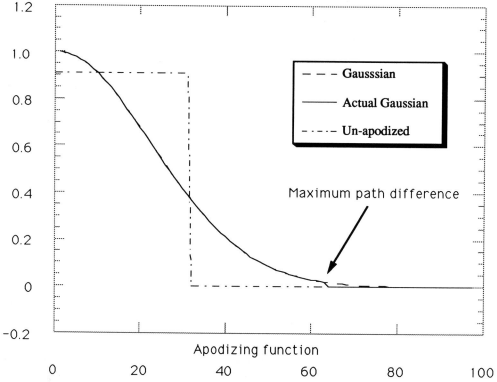

Figure 3. The apodization functions used for Figs 1 and 2. The Gaussian is just truncated at maximum path difference. This is the reason for the lobes in the response in Fig. 2.

3.2. Resolution

The resolution of the Michelson is variable. It depends only on the maximum path difference of the interferogram. The maximum path difference is set by the design of the instrument but within this bound the resolution used for a particular observation depends only on where the instrument is scanned to, which is a matter of software and user choice. The number of points in the final spectrum is the same as the number of points observed in the interferogram. The resolution of an F-P system is fixed by the design of the instrument.

3.3. Efficiency of light usage

The Michelson uses all of the light. The only light not measured is that lost due to instrumental absorption or scattering. The modulation efficiency of a Michelson is given by $4RT$ where R is the reflectivity of the beam-splitter, and T is its transmission. This can be made to be 90% fairly easily. The only real problem is differential polarization between reflected and transmitted beams which can have a greater effect on overall efficiency than intensity imbalances. If both output ports are used the instrument switches the input light from one output port to the other. It is possible to use an F–P in this way by taking the light reflected from the back of the F–P. This, however, means using it off-axis with the attendant problems. It is not possible to use two F–Ps in series in this way.

3.4.   Scintillation

A consequence of the above is that the Michelson is insensitive to intensity scintillation. The fringe depth can be obtained from the difference between output ports over their sum. This gives an amplitude rejection ratio of between 10:1 and 100:1. The only effective way to reduce scintillation noise with an F–P is to scan rapidly.

3.5.   Background subtraction and the alternative input port

The alternative input port of the Michelson allows the input port to be switched making a completely symmetrical system. The instrumental effects are cancelled to first order. By using this technique it is possible to produce a system with a stability of a fifty-thousandth of a fringe [Forrest, 1986]. This is not necessary in most systems. In the infra-red the alternative input port can be used very conveniently for sky subtraction. By comparing the output of the interferometer from just sky, and then sky plus source, accurate sky subtraction can be obtained. It is also physically convenient to do this because the alternative input port can be fed from a piece of image close or far from the source of interest by moving one mirror (Fig. 4). The optimum distance depends on source and wavelength.

3.6.   Calibration accuracy

The calibration accuracy of the Michelson is extremely good. The moving mirror is servo-controlled against a frequency standard, usually a laser. For noise considerations a stability of about one hundredth of a fringe is necessary. This has the side-effect of making the sampling position in the interferogram good to near this accuracy. The line centre information from the instrument can therefore be much more accurate than the nominal resolution of the interferometer.

3.7.   Construction

The interferometer is relatively simple to build. There is essentially one moving part, a mirror. For a resolution of 2000 the movement is so small that spring couplings can be used removing all sliding and rolling joints. The positional accuracy of the mirror is not as demanding as in the F–P because the interferometric beam is not folded (or at most only once). The movement, however, must be longer by the same

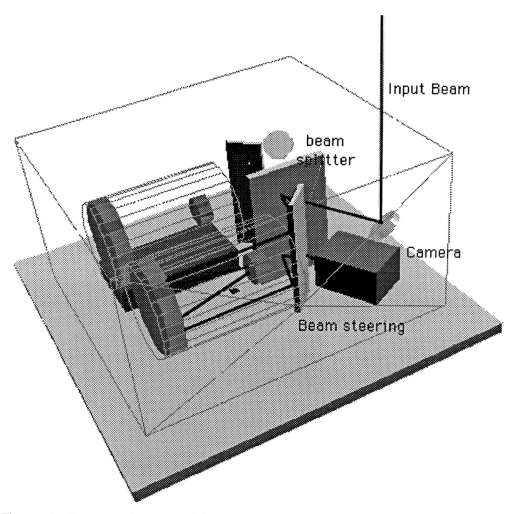

Figure 4. Layout of proposed instrument; oblique view of instrument showing steering mirrors and folded path. It will be mounted at the Cassegrain focus.

factor. This compares with the case of twin F–Ps where parallelism as well as gap must be controlled giving a minimum of four coupled servo-systems.

4.   The traditional view of Michelsons

The usual perception of the Michelson is that it has very high resolution and is very difficult to build and operate. The data need weeks of reduction before being understood. The net result of this view is that Michelsons are used only for these very difficult high-resolution problems for which the Michelson is the only choice. This, compounded with the unfamiliarity of the Michelson, re-enforces the idea that these are difficult specialist machines. The ultra-violet Michelson of R. Learner and

L. A. Thorne is a case in point. The task the instrument is used for is almost impossible, but very good results have been obtained. However, the Michelson has some very real advantages even when the 'special' advantages do not apply. In particular Michelsons work just as well at low resolution as at high resolution.

5. Multiplex advantage

The multiplex or Felgett advantage is not necessary for the Michelson to compete in terms of signal-to-noise. For the case of noiseless detectors the Michelson and F–P give almost identical signal-to-noise for similar observations of a pre-filter passband. There are special cases which give one or other an advantage.

5.1. Field widening

Field widening allows a Michelson to have even smaller optics at the cost of some complexity. Even without field-widening, however, the area of the beam in the Michelson will be about the same as in the F–P. The Jaquinot criterion [Jacquinot, 1954] can be broken for both the Michelson and F–P by using computing techniques to decode the data. This has been well demonstrated in the Taurus instrument [Atherton *et al.*, 1982] [Atherton, 1983]. Field widening therefore need not be used if the resolution is fairly low. The limiting factor for an imaging Michelson or F–P is in fact the dispersion across one pixel at the edge of the field of view.

5.2. Traditional disadvantages

The Michelson also has some traditional disadvantages. First, it is difficult to build. This as mentioned is because most of the instruments built have attempted to solve extremely difficult problems. Low-resolution instruments are relatively easy to build.

The second disadvantage is that they are difficult to use. Scanning the interferometer *is* more difficult than looking at different points on an output spectrum. Modern astronomy already has a multitude of control and servo-systems, most of which an observer takes little notice of. There is no reason why the control system for a Michelson can not be made as easy to use and unintimidating as other parts of the telescope control.

Lastly the data take a large effort to reduce. With modern computers such as the transputer it is entirely possible, at reasonable cost, to Fourier-transform the data during the observation, even for an imaging Michelson. The astronomer would take away data already in the wavelength domain.

6. The proposed instrument

We are trying to get support to solve some of the problems of producing a common user, imaging, low-resolution spectrometer for use on 8-m telescopes. The proposed characteristics of the instrument are :

1. 90 arcsec field of view on a 8-m telescope.

2. Wavelength ranges: 500–900 nm (visible configuration) 1–5 μm (IR configuration). The instrument is the same for the entire wavelength range apart from the beam-splitter and detectors.

3. Maximum path difference of 10 mm $\rightarrow \mathscr{R}_{max} = 10^4$ at $\lambda = 1\mu$m, $\mathscr{R}_{max} = 2000$ at $\lambda = 5\mu$m.

4. Two input and two output ports for background subtraction or sky transparency compensation.

5. All-aluminium alloy construction, apart from optically-transmitting components. The mirrors would be diamond-turned aluminium alloy. This maintains good alignment stability on cooling.

6. f/15 input f/4 output $\rightarrow$ 0.25 arcsec pixels in the visible, 1 arcsec pixels in the IR when using SBRC 58 × 62 arrays which would be used for the test instrument.

6.1. Limitations

The final performance of the instrument will largely be determined by the characteristics of the CCD or diode array detectors. The prototype instrument will operate in the visible to start with, but will be fabricated from parts which will allow it to be cooled when IR detectors are available. The noise of the CCD is limited by readout noise. This produces an effect a little like Felgett or multiplex advantage.

The difference is that the noise is independent of exposure time. This implies that the CCD should be read out as few times as possible. This in turn implies a step and integrate rather than a continuous scanning mode of operation. This is further required by the fact that current large CCDs take a significant time to read out. To get maximum signal-to-noise with current CCDs, multiple readout without clearing the chip is used. This can mean readout taking up to a second. The diode array has fewer readout problems associated with it.

There is a multiplex disadvantage. Although the signal on the detector is larger than the equivalent bandpass spectrometer the signal-to-noise required from each measurement is greater. These two effects exactly balance for shot noise, at least in theory, but more dynamic range is required of the interferometer detector. The linearity of the detector must therefore be very good, and better than the equivalent F–P.

6.2. Computing

A 512 × 512 CCD will require a quarter of a million Fourier transforms per observation. The best way to perform these currently is with the i860 computer chip which will do a thousand-point transform in ~ 1 ms. Several of these may be required to do the Fourier transforms during the next observation. There are several other computing tasks required. The output CCDs cannot be aligned to better than a pixel; the images should therefore be re-sampled to obtain optimum spatial resolution. The Michelson servo-system needs to have very high performance so that the time spent moving between observing points on the interferogram is negligible and that the stability while integrating is very good. A computer-driven servo is the best way to achieve this. The instrument also requires general computer control and the resulting data must be stored rapidly.

The obvious way to accomplish these parallel tasks is to use a parallel computer network. An array of Inmos transputers with i860s as signal-processing chips and a direct SCSI interface will do all the above tasks and allow very great flexibility.

There will be a large amount of data generated by such a system. Recently this has become much less of a problem; 600 Mbyte can be put on one CD ROM disc at quite low cost. The data on these discs can be examined even on small desktop systems such as a Macintosh with large screen, and costing much less than £10 000. The data will already be in the wavelength domain so much simple data reduction will not need special facilities.

7.   Conclusion

It appears possible to produce an imaging spectrometer producing as detailed images as any current broadband electronic imaging and a spectral resolution of 2000 or better at any wavelength between 0.5 μm and 5.0 μm. The amount of data produced is large but can be handled with modern computing. If built, the flood of data produced from such an instrument on a large telescope should encourage a burst of new astronomy.

References

[Atherton, 1983] Atherton P.D., in *Instrumentation in Astronomy V. SPIE Vol. 445 (1983), p. 535.*

[Atherton *et al.*, 1982] Atherton P.D., Taylor K., Pike C.D., Harmer C.F.W., Parker N.M., Hook R.N., *Mon. Not. R. astr. Soc., Vol. 201 (1982), p. 661.*

[Forrest, 1986] Forrest A.K., in *Interferometric Stellar Oscillation Spectrometry: NATO ASI series C, Mathematical and Physical Science. No. 169 (1986), p. 391.*

[Jacquinot, 1954] Jacquinot P., *J. Opt. Soc. Am., Vol. 44 (1954), p. 761.*

Additional references for background:

[Connes & Michel, 1975] Connes P., Michel G., *Applied Optics, Vol. 14 (1975), p. 2087.*

[Fellgett, 1951] Fellgett P., *PhD Thesis, Cambridge University. (1951).*

[Hall, 1986] Hall D., *Imaging Michelson Interferometer for Space Telescope. U. of Hawaii Institute for Astronomy (1986).*

[Huppi *et al.*, 1979] Huppi R.J., Shipley R.B., Huppi E.R., *Multiplex and/or High Throughput Spectroscopy; SPIE, Vol. 191 (1979), p. 26.*

[Maillard, 1988] Maillard J.P., in *The impact of Very High S/N Spectroscopy on Stellar Physics; Proc. IAU Symp. 132 (1988), p. 71.*

[Maillard & Michel, 1982] Maillard J.P., Michel G., in *Instrumentation for Astronomy with Large Optical Telescopes; Proc. IAU Colloq. 67 (1982), p. 213.*

[Steed, 1979] Steed A.J., *Multiplex and/or High Throughput Spectroscopy; SPIE, Vol. 191 (1979), p. 2.*

12

Three-Mirror Telescope – Optical Tests and Observations

R.V. Willstrop *

Abstract

The history of the Three-Mirror Telescope design is described briefly. The performance of a prototype of 0.5-m aperture is demonstrated by some photographs taken with it. Methods of testing each of the three mirrors during construction, both of the prototype and of a hypothetical telescope 5 m in diameter, are described.

1. Introduction

About eight years ago Professor Donald Lynden-Bell asked me to do some optical design work to see if it was possible to make a Schmidt with a large aperture and a short tube. This proved to be impossible, but later I followed a suggestion by Dr. E. J. Kibblewhite that I should take a look at the Paul–Baker system. This had been published first by Maurice Paul [Paul, 1935], was rediscovered independently and mentioned briefly by James Baker [Dimitroff & Baker, 1945], and was described in detail as a corrector system for the Hale 200-in telescope [Baker, 1969]. I found there was a wide-field all-reflection system which might replace the Schmidt in large sizes [Willstrop, 1984; Willstrop, 1985].

The purpose of such a telescope would be two-fold: direct photography, and fibre-optic spectrography of many faint objects simultaneously, as with the FLAIR system on the UK Schmidt,[1] but reaching fainter objects because of the larger aperture.

There is a family of these designs, with field of view increasing from 5 deg at f/1.6 to 8 deg at f/1.0. The ray-theoretical worst image spread increases from 0.33 arcsec at f/1.6 to 3.6 arcsec at f/1.0, so the f/1.6 version is the best of these for a large ground-based telescope.

Bernhard Schmidt knew that the practical limit to the aperture of his camera

---

*Institute of Astronomy, Madingley Road, Cambridge CB3 0HA, UK.
[1]See Watson, this volume. – *Ed.*

139

was about 48 in [Baade, 1953]. Perhaps the Palomar Schmidt was made this size on Baade's advice. An f/2.5 spherical mirror without any corrector forms images with about 100 arcsec of spherical aberration, and a corrector made of a single piece of crown glass can reduce this to about 1 arcsec over a narrow region of the blue and near ultraviolet spectrum. With a focal length of 3 m, this image spread was an acceptable match to the grain of the fastest emulsions available in the 1930s.

Astronomers now wish to observe over a much wider spectral range, from near the atmospheric limit in the ultraviolet to about 1 μm in the infrared. At the ends of this range a singlet corrector gives image spreads of 3 to 4 arcsec. The singlet aspheric correctors on the UK, Palomar and ESO Schmidts have now been replaced by doublets; the first of these was designed by Charles Wynne, whose 80th birthday we are now celebrating [Wynne, 1981]. With a doublet the chromatic variation of spherical aberration is much reduced, but because of irrational dispersion it cannot be eliminated; in the middle of the visible spectrum the aberration is slightly over-corrected, and at the ends of the spectrum it is under-corrected. The best that can be done with 'normal' glasses, such as Schott BK7 and LLF6, is to eliminate it at two wavelengths, such as 400 and 750 nm, and then the axial images will be smaller than 0.25 arcsec from 350 to 880 nm. These images provide an acceptable match to modern fine-grain emulsions such as IIIa-J and IIIa-F, again at a focal length of about 3 m.

The main advantages of the three-mirror design, compared with the Schmidt, are that it is perfectly achromatic, and it can, in principle, be built with a much larger aperture, both because there is no question of having to obtain large pieces of glass for a corrector, and also because it is much more compact. The separation of the second and third mirrors is equal to the focal length, while the corrector plate and spherical mirror of the Schmidt must be separated by twice the focal length (Fig. 1). Its other advantages are: a very much wider field of view than any other all-reflection telescope, and almost as large as the Schmidt; the image quality is better than in the Schmidt when it is built on a large scale to minimise the effects of diffraction; there are no Schmidt-type ghost images if it is used without a filter, and only very faint ghost images if a filter is used; baffles can be arranged to prevent light from the sky reaching the focal surface without first being reflected in all three mirrors; and the image spread can be kept smaller than 1 arcsec over the full 5-deg field when an atmospheric dispersion compensator is added if both mirror separations (1-2 and 2-3) are adjusted. The Schmidt does not allow this [Willstrop, 1987].

There are some disadvantages: the illumination is uniform only over the central 2 deg, and vignetting increases to 22% at 2.5 deg off axis; the focal surface is slightly less accessible than in the Schmidt, because it is surrounded by the primary mirror as well as by the incoming light; and the tube is not closed as in the Schmidt so the mirrors, two of them face up, do not stay clean for so long.

So far, two examples of this telescope have been built here, a working model of

140

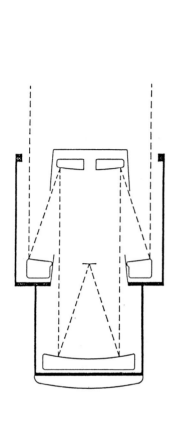
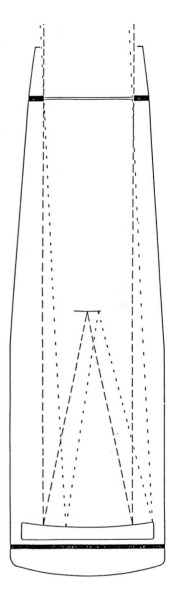

Figure 1. The paths of light rays through the Three-Mirror Telescope (left) and Schmidt telescope (right). As drawn, these two telescopes have equal light-collecting areas; note the short tube of the Three-Mirror Telescope.

100 mm aperture, and a prototype of 0.5 m aperture which is in a wooden building to the south of the Solar Physics Observatory. The mirrors of the prototype were made spherical at A.E. Optics in Dry Drayton, 5 miles from Cambridge; I took responsibility for the aspheric working.

141

Figure 2. A photograph of the nebula IC443 and ηGem as imaged with the prototype 0.5-m three-mirror telescope on 8 March 1991; the image is $\sim 64 \times 44$ arcmin.

2. Observations

More than 100 photographs have been taken with the 0.5-m prototype since it was installed during August 1989. Some of these have been focus tests, or for polar axis alignment, or to test the smoothness of the sidereal drive or the flexure of the guide telescope. There is space here to show only a photograph of IC443 (Fig. 2).

The best images are 3 or 4 arcsec in diameter, which is 10 or 12 times the ray-theoretical image size. However, with this relatively small prototype, diffraction

142

causes image spread of about 1.6 arcsec. Focus-test photographs with images on each side of the best focus reveal a little astigmatism. This must have arisen during my aspheric working of the primary mirror, because it is not possible to introduce it by misalignment of the mirrors without also causing very much larger amounts of coma, which in fact is absent. Accuracy of focusing is slightly variable, as if the film holder support is flexing under gravity. The sidereal drive is good enough for exposures up to about 15 or possibly 20 min, but trailing occurs in longer exposures. Similarly, the Celestron guide telescope works well in short exposures, but there is a little flexure between it and the main telescope if the exposure is longer. A fibre-optic image guide has been ordered, and this will form part of a guiding system using a star image just outside the film holder.

The conclusion is that all of the defects of the prototype are relatively simple mechanical problems, and that the optical design is not at fault. A large telescope would be properly engineered, not designed on the backs of envelopes (or on computer line-printer paper as I did this one). Improved figuring techniques, such as those I used on the second and third mirrors, would give the primary mirror a smoother optical surface, and a small amount of astigmatism could be corrected with the type of mirror supports used in the ESO New Technology Telescope (NTT). There would then be no problems in obtaining good performance.

3.   Optical tests

The choice of optical test for each mirror during figuring depends on several factors; is the mirror convex or concave, is it aspheric or not, what is its aperture, and what accuracy is needed? A concave spherical mirror is the most convenient to test, because if a light source is placed close to its centre of curvature it will form a diffraction-limited image. This can be tested with a knife edge, Ronchi grating, interferometer or eyepiece. The interpretation of the test is made more uncertain, so that unacceptable surface errors might remain undetected, if the image is not diffraction-limited. For ellipsoids, the light source and knife edge must be moved along the optical axis to the conjugate foci. For all other types of mirror, auxiliary optics are needed to create a null test. For example, in the Ritchey test, a concave paraboloid can be tested in auto-collimation with a flat mirror, which should be as large as the paraboloid [Ritchey, 1904]. The accuracy of any test depends on the quality of the auxiliary optics used, and its cost depends on the size of any new components that must be made.

In general, therefore, existing mirrors or lenses of known quality should be used whenever this is practicable; and if new auxiliary optics must be made the test should be designed so that they are not unnecessarily expensive. I will describe optical tests for the mirrors of the prototype and of a hypothetical telescope with a primary mirror 5 m in diameter.

Table 1. Null test for primary mirror of 0.5 m aperture

	Radius (mm)	Separation (mm)	Aperture (mm)
Source			
		2928.5	
Mirror	−6093.0		493.2 illuminated
		2500.0	
Mirror	−1600.0[1]		509.0
		808.364[2]	
Flat	Plane		1.0 illuminated
		808.364[2]	
Mirror	−1600.0[1]		509.0
		2500.0	
Mirror	−6093.0		493.0
		2928.5	
Focus[3]			

[1]Axial radius of curvature of the mirror under test; the power series defining its profile is given in Willstrop [1987]. This mirror has a perforation 300 mm in diameter.

[2]The axial distance of the small flat from the primary mirror does not need to be measured; the flat is correctly positioned when the final image coincides with the source. An error of 1 mm in the distance of the source or the final image from the spherical mirror causes the rms optical path variations to be increased by approximately 1%, 2 mm causes an increase of 4%. The accuracy of the test is very insensitive to changes in the separation of the concave mirrors.

[3]The rms optical path variation at focus is 49 nm in this double-pass test.

3.1. Primary mirror

The primary mirror of the 0.5-m prototype differs significantly from the paraboloidal form, so the Ritchey autocollimation test with a flat mirror is not suitable. Fortunately A.E. Optics has a 25-in concave spherical mirror which could be used in a modification of a test first described by Tom Waineo [Waineo, 1981]. Light from a small source shines through the hole in the primary mirror and is made nearly parallel by the sphere, though still diverging slowly. The light is then focused by the primary mirror. This image is inaccessible, so in this position I placed a small flat mirror, which returns the light (reversed top to bottom and left to right) to the primary and the sphere back to the source (Table 1); a beam splitter is needed in front of the source. This test is free of coma because of the inversion. The rms optical path variation is 49 nm in the double pass test (24.5 nm at each reflection

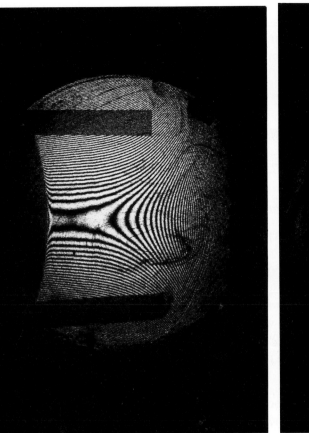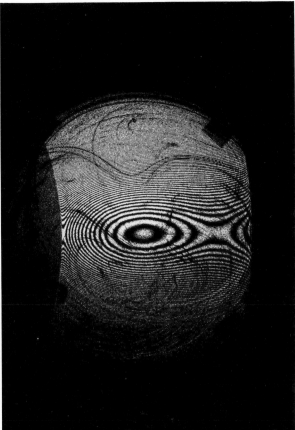

Figure 3. Interference tests of the primary mirror with two convex proof spheres of slightly different radii. Note the irregularities in the fringes at one-quarter of the distance from the inner to outer edge.

in the primary mirror). The departure of the ideal surface from a paraboloid intersecting it at radii of 150 and 255 mm is 97 nm (rms) or ± 150 nm (peak to peak). A knife-edge test showed many narrow zones, which had rather high contrast because of the two reflections in the mirror. Interference tests using convex proof spheres (Fig. 3) show that the height of the worst zones is only about 0.2 fringes of He–Ne laser light (63 nm on the mirror surface). I used a better technique on the secondary mirror which produced a much smoother surface. Calculations of fringe positions confirm that the overall shape is correct. The Waineo test is both impractical and too inaccurate for a primary mirror 5 m in diameter. In this case the Offner null test [Offner, 1963] is to be preferred. Offner needed to figure a 36-in f/4 mirror to high accuracy for the Stratoscope. Primary spherical aberration (3rd order) is corrected by a plano-convex lens. A small plano-convex field lens having the correct

Table 2. Null test for primary mirror of 5.0 m aperture

	Radius (mm)	Separation (mm)	Index	Aperture (mm)
Source				
		502.6727	1.0 (air)	
	Plane			181.84
Lens 1		50.0	1.51759	
	−160.0			186.57
		667.5157	1.0	
	−5326.242			98.83
Lens 2		12.7	1.51759	
	−517.7507			101.68
		15 857.3	1.0	
Mirror	−16 000.0[1]			5000.0[2]
		16 200.0	1.0	
Focus[3]				

[1]Axial curvature of mirror, defined by power series.
[2]The central perforation of the mirror is 2969 mm in diameter.
[3]The rms optical path variation at the focus is 0.154 nm in this single-pass test.

power, and placed at the intermediate focus, will remove 5th order aberration or leave the correct amount of this to balance uncorrected higher orders. The Offner test has been widely used for many large mirrors up to that of the William Herschel Telescope (4.2 m, f/2.5).

The residual errors of the test increase in proportion to the aperture and very rapidly with the focal ratio, so that for an f/1.6 mirror it is worth trying to make some improvement. Changing the shape of the larger lens makes a gross change in its properties, but does not appear to reduce the optical path residuals. It occurred to me to try the effect of moving the field lens along the axis, away from the circle of minimum confusion, and this made some improvement. Then David Brown of Grubb Parsons suggested (private communication) that if the field lens was given a meniscus or biconvex shape, instead of being strictly plano-convex, the optical path residuals might be reduced still further.

This is correct; there are two positions of the field lens which give very small optical path residuals, 0.154 and 0.493 nm rms if the field lens also has the optimum shape (Tables 2 and 3). In these two arrangements not only are the 3rd and 5th-order spherical aberrations corrected, but also the 7th and 9th. The same plano-convex compensator lens is used in each. These residuals are very much smaller than would be caused by typical variations of the refractive index of the best optical glass.

Table 3. Alternative null test for the primary mirror of a 5.0 m telescope

	Radius (mm)	Separation (mm)	Index	Aperture (mm)
Source				
		306.3882	1.0 (air)	
	Plane			202.01
Lens 1		50.0	1.51759	
	−160.0			207.06
		536.24	1.0	
	653.11			87.83
Lens 2		12.7	1.51759	
	−1350.1			85.91
		16 459.0	1.0	
Mirror	−16000.0[1]			5000.0[2]
		16 200.0	1.0	
Focus[3]				

[1] Axial curvature of mirror, defined by power series.
[2] The central perforation of the mirror is 2969 mm in diameter.
[3] The rms optical path variation at the focus is 0.493 nm in single-pass.

Any slab of glass which is to be made into such a lens ought first to be polished nearly flat on all six faces and tested interferometrically. The alternative is a mirror compensator, as was used for the Hubble Space Telescope primary mirror.

3.2. Tertiary mirror

I shall depart from the apparently logical order, and discuss the testing of the tertiary mirror next because this may be used in testing the secondary, and like the primary, it is concave.

The third mirror should be very nearly spherical, with a slightly high edge, the asphericity increasing mainly with the 6th power of the distance from the axis. This is opposite to the sign of the asphericity of a paraboloid, which increases with the 4th power of the distance off axis. So the Offner test will not work.

In the 0.5-m prototype the asphericity reached $0.823\,\mu$m at 190 mm from its axis, relative to the sphere of 800 mm radius which is the best fit near the axis. To produce this shape it would be necessary to lower the greater part of the surface, uniformly, by this amount. Instead, I chose to refer the figure to a slightly shorter radius, 799.972 mm. In this case it is only necessary to polish away a broad groove $0.41\,\mu$m deep. This was done in one afternoon, in a total of just over one hour polishing. Testing was done with a convex proof sphere, 150 mm in diameter. This

147

Table 4. Null test for the tertiary mirror of a 5.0 m telescope

	Radius (mm)	Separation (mm)	Index	Aperture (mm)
Source				
		7940.0	1.0	
Mirror 3	−8000.0[1]			3660.0
		7920.0	1.0	
	−120.0			67.10
Lens 1				
		10.0	1.517 59	
	Plane			65.75
		372.658	1.0	
	87.854			13.98
Lens 2				
		8.0	1.517 59	
	552.0			12.85
		243.4975	1.0	
	40.0			39.35
Lens 3				
		16.0	1.517 59	
	Plane			36.69
		97.618	1.0	
Focus[2]				

[1]Axial radius of curvature of the mirror defined by power series.
[2]The rms optical path variation at focus is 0.353 nm in single pass.

had a slightly defective edge, so only the central 5 in (127 mm) was used. The proof sphere was used in three positions, centred to check up to 63.5 mm, from the axis to 127 mm, and from 63 to 190 mm. In the first two positions the proof sphere should show practically the same fringe pattern, but in the third position it should show about 1 fringe extra concavity. Polishing was continued until this was reached. It is not an ideal test by any standard, but at the time I had not been able to arrange any better. I will return to the prototype after the 5 m version.

It was noted above that the third mirror has a turned-up edge; this causes the focus of the edge rays to be closer than that of paraxial rays to the mirror. If a plano-concave lens is placed in the converging light the focus of the edge rays can be lengthened more than the paraxial rays, so that the edge rays again come to the longer focus. Then, with two more lenses arranged as in an Offner compensator, the light from all zones can be brought to a good focus (Table 4). The plano-concave lens will not produce the desired effect if it is used alone. With the radii of −120 and +40 mm predetermined, the conjugate foci of the third lens can be chosen to correct the 3rd order spherical aberration, and the power, position and shape of

Table 5. Null test for the tertiary mirror of the 0.5 m prototype

	Radius (mm)	Separation (mm)	Index	Aperture (mm)
Retro-reflecting Flat				
		764.1	1.0	
Mirror 3	−800.0[1]			381.0
		765.6	1.0	
	Plane			34.31
Lens 1		3.4	1.51769	
	52.2			32.50
		458.25	1.0	
	52.2			24.99
Lens 2		8.5	1.52257	
	Plane			24.38
		183.75	1.0	
Source and focus[2]				

[1]Axial radius of curvature of the mirror, defined by power series.
[2]The rms optical path variation at focus is 7.4 nm in single pass or 15 nm in double pass.

the field lens can all be adjusted simultaneously to reduce the rms path difference to 0.353 nm. Again, this is an academic figure; slight inhomogeneities in the lenses would affect the result. This is a single-pass test, and the mirror is slightly tilted to bring the image to a convenient position.

Returning to the prototype, I found that if the apertures of the two outer lenses are in the appropriate ratio, then the power of the field lens is zero. It can therefore be removed. We can only correct the 3rd and 5th order aberration, but that is acceptable in this smaller-scale test. It would be necessary to tilt the mirror significantly to bring the focus clear of the concave lens and its mounting, so a small (2 mm diameter) flat mirror was secured to the lens with Bluetack to make it a double-pass test (Table 5). A beam splitter is again needed near the source. The rms path variation in the double-pass test is 14.8 nm, or 7.4 nm at each reflection on the tertiary mirror.

3.3. Secondary mirror

We can now return to the secondary mirror. Because this is convex, it cannot focus light if it is used alone. Any null test is therefore more complex than for a concave mirror. Until about 95% of the aspheric work had been completed, testing was by interference with a laser source and a concave proof sphere. The

149

mirror is made of Zerodur, which is a brown colour and slightly diffusing, but in a thickness of only 2 in and with red laser light it is quite clear enough. The proof sphere was brought up close to the mirror, with a piece of graph paper between them to avoid contact, and to indicate the scale. The fringe separation can be adjusted by tilting the proof sphere. In Fig. 4a the air gap is constant along a radius, up to about 60 mm off axis. There is a turned-down edge on the proof sphere, but otherwise the fringes run straight and smooth over this range. Beyond about 90 mm the fringes are too close together to give a sensitive test, but if the proof sphere is now tilted about an axis at right angles to the fringes, then the hump in the fringes rolls smoothly out to the edge of the mirror (Figs. 4b–4f). A second proof sphere with a radius of 795.5 mm shows that the outer zones are smooth (Fig. 4g). The total asphericity when these photographs were taken was about 40 fringes before meeting a steep turned-down edge covering the last 3 mm. The mirror has since been edged down from 305 to 280 mm in diameter. The secondary mirror is much smoother than the primary because I used a flexible polisher designed to polish away the high zones, averaged over a distance of a few cm. This polisher was moved radially over the mirror, by hand, making 20 to 30 strokes in each revolution of the mirror. The primary had been worked entirely by machine, with relatively slow radial motion of the polishers.

These interference tests show that the mirror is smooth, but unless we can estimate the position of the fringes with an accuracy better than one tenth of their spacing, or about 1 mm at a radius of 100 mm and 0.1 mm at the edge of the mirror, we can say nothing about the overall shape. Again an optical null test is needed.

The Hindle test [Hindle, 1931] for a convex hyperboloidal mirror, for example the secondary of a Cassegrain telescope, uses a light source at one conjugate focus to illuminate the convex, and a concave spherical mirror with its centre of curvature coincident with the other, inaccessible, conjugate focus, to return the light to the source. For large convex secondaries, a meniscus lens can be used instead of a spherical mirror. A sphere is a special case, with both of its conjugate foci coincident and inaccessible. So I had to use a little lateral thinking - the Cassegrain

Figure 4. Interference tests of the secondary mirror (300 mm in diameter) with a concave proof sphere (150 mm in diameter). The small dark ring at the left of each section is a fiducial mark at the centre of the secondary mirror. (a) Uniform air space along one radius (horizontal in this figure) up to about 60 mm from the centre of the secondary; (b) Proof sphere tilted 0.3×10^{-4} rad; (c) Proof sphere tilted 1.0×10^{-4} rad; (d) Proof sphere tilted 1.5×10^{-4} rad; (e) Proof sphere tilted 2.5×10^{-4} rad; (f) Proof sphere tilted 4.2×10^{-4} rad; (g) Test with a second proof sphere of slightly shorter radius.

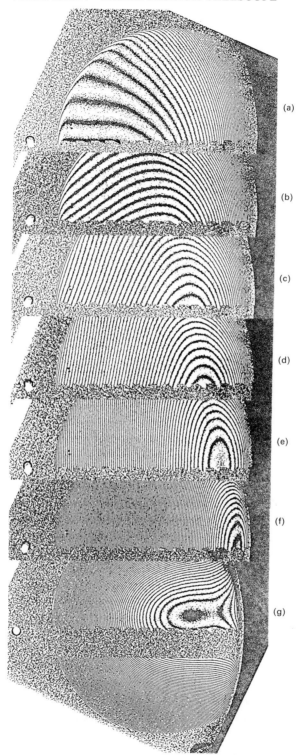

Figure 4. For legend see opposite

151

Table 6. Null test for the secondary mirror of the 0.5 m prototype

	Radius (mm)	Separation (mm)	Index	Aperture (mm)
Source				
		328.6175	1.0	
Mirror	−697.3			255.40
		600.0	1.0	
	409.35			273.50
Meniscus lens		18.161	1.517 59	
	400.0			269.54
		25.78	1.0	
Mirror 2	800.0[1]			270.0
		25.78	1.0	
Lens (M)	−400.0			279.04
		25.78	1.0	
Mirror 2	800.0[1]			270.0
		25.78	1.0	
	−400.0			269.54
Meniscus lens		18.161	1.517 59	
	−409.35			273.50
		600.0	1.0	
Mirror	−697.3			255.40
		328.62		
Focus[2]				

[1] Axial radius of curvature of the secondary mirror, defined by power series.
[2] The rms optical path variation at focus is 38 nm in double pass or 19 nm in single pass.

hyperboloidal secondary is tested with light falling on it at the same angles of incidence as in the telescope, so it should be possible to test this nearly spherical secondary mirror in an equivalent way.

Light from a star on the axis of the Three-Mirror Telescope is reflected by the primary mirror to a virtual image 8 arcsec in diameter, and the final axial image is under 0.1 arcsec (ray-theoretical image spread). Therefore a source placed at the final focus, shining on the third mirror, and reflected by the secondary, will appear to come from a small virtual image at the focal point of the primary. If a meniscus lens is placed with its centre of curvature at this position, some of the light will be reflected back via the two mirrors to an image close to the source.

The passage of the light twice through a meniscus lens with concentric surfaces

Table 7. Null test for the secondary mirror of a 5.0 m telescope

	Radius (mm)	Separation (mm)	Index	Aperture (mm)
Source				
		3750.0	1.0	
Mirror 3	-8000.0^{1}			2586.2
		4600.0	1.0	
	4200.0			2739.5
Meniscus lens		200.0	1.51759	
	4000.0			2696.3
		200.0	1.0	
Mirror 2	8000.0^{1}			2700.0
		200.0	1.0	
Lens (M)	-4000.0			2753.5
		200.0	1.0	
Mirror 2	8000.0^{1}			2703.4
		200.0	1.0	
	-4000.0			2703.4
Meniscus lens		200.0	1.51759	
	-4200.0			2752.6
		4600.0	1.0	
Mirror 3	-8000.0^{1}			2811.4
		3949.314	1.0	
	104.9656			43.26
Field lens		10.0	1.51759	
	518.3			38.26
		88.3185	1.0	
	50.0			51.69
Compensator lens		20.0	1.51759	
	-160.0			52.50
		921.82	1.0	
Focus^{2}				

[1]Axial radii of mirrors defined by power series.
[2]The rms optical path variation at focus is 6 nm in double-pass test.

introduces both chromatic and spherical aberration. Longitudinal chromatic aberration was corrected by making the lens radii differ by only 9 mm although the thickness was 18 mm. Spherical aberration could be corrected by using a miniature Offner compensator in front of the source, but in the prototype the lenses would

153

be inconveniently small. In place of the third mirror, with a radius of −800 mm, I therefore made a new concave mirror, 300 mm in diameter and with a radius of curvature of −697.3 mm, specially for this test (Table 6).

It is difficult to obtain a disc of glass large enough to make a meniscus test lens for the secondary mirrors of the largest telescopes now under discussion, and David Brown suggested that I should make this lens in four pieces. It was necessary always to work the thickest one, and to keep the correct wedge angle, to end up with four identical lenses, but there was no great difficulty. In the test arrangement the four lenses were mounted on simple screw adjustments and an attempt was made to bring their centres of curvature into coincidence on the axis of the system. This was not entirely successful; it was never possible to obtain a good knife-edge test through more than one of the lenses at a time. However, in the Keck telescope 36 segments are being adjusted; it is not impossible.

The knife-edge test revealed some faint zones with very low contrast, showing that the shape of the secondary mirror is close to the ideal.

For a 5-m version it would be costly to have to make a concave mirror more than 2.5 m in diameter specially for this test, so the third mirror should be used. It can be in its final form, with the turned-up edge, and the aberration caused by the meniscus lens can be corrected with a small Offner corrector (Table 7). The tolerances on the radii of these lenses are not unreasonably tight. Good engineering methods will result in optical-path errors of the order of $1\,\mu\text{m}$ rms on the mirror surface; with readjustment of the separation of the mirrors in the completed telescope the ray-theoretical image spread will be increased by not more than 0.5%.

4.   Conclusion

The building of a large telescope of this type is entirely practical.

References

[Baade, 1953] Baade W., in *Amateur Telescope Making, Book 3.* ed. Ingalls, A.G., Kingsport, Tennessee: Scientific American, Inc., 1953, p. 371.

[Baker, 1969] Baker J.G., *IEEE Trans., Vol. AES-5 (1969), p. 261.*

[Dimitroff & Baker, 1945] Dimitroff G.Z., Baker J.G., *Telescopes and Accessories.* Philadelphia: Blakiston, 1945, p. 105.

[Hindle, 1931] Hindle J.H., *Mon. Not. R. astr. Soc., Vol. 91 (1931), p. 592.*

[Offner, 1963] Offner A., *Applied Optics, Vol. 2 (1963), p. 153.*

[Paul, 1935] Paul M., *Rev. Opt. Theor. Instrum., Vol. 14 (1935), p. 169.*

[Ritchey, 1904] Ritchey G.W., *Astrophys. J., Vol. 19 (1904), p. 53.*

[Waineo, 1981] Waineo T., *Telescope Making, No. 11 (1981), p. 12.*

[Willstrop, 1984] Willstrop R.V., *Mon. Not. R. astr. Soc., Vol. 210 (1984), p. 597.*

[Willstrop, 1985] Willstrop R.V., *Mon. Not. R. astr. Soc., Vol. 216 (1985), p. 411.*

[Willstrop, 1987] Willstrop R.V., *Mon. Not. R. astr. Soc., Vol. 225 (1987), p. 187.*

[Wynne, 1981] Wynne C.G., *Q. J. R. astr. Soc., Vol. 22 (1981), p. 146.*

IV Maximising resolution

13

Bispectral Imaging: Past, Present and Future

J. C. Dainty*, A. Glindemann* and R. G. Lane *

Abstract

Bispectral imaging is a modification of the technique of speckle interferometry that allows diffraction-limited images to be obtained from ground-based large astronomical telescopes. In this paper, we review the basic principles of the method, give some examples of its application to astronomical imaging and speculate on future developments.

1. Introduction

The modern era of high angular resolution imaging using single large Earth-based telescopes commenced in 1970 with the invention by A. Labeyrie of the technique of speckle interferometry [Labeyrie, 1970]. The influence of this invention and related work by Labeyrie on the development of high angular resolution imaging has been very considerable. This is probably because, in speckle interferometry, the issue of the degradation caused by atmospheric turbulence or "seeing" has been tackled head-on — in fact, in speckle interferometry one *uses* the "seeing" to give diffraction-limited resolution, even when the telescope optics are imperfect. As a result, there are now a whole host of techniques, from simple image sharpening to adaptive optics for overcoming the deleterious effects of "seeing".

   Speckle interferometry itself has had, until recently, limited success and is still not accepted as a routine observing technique by the majority of observational astronomers in the visible and near infra-red. This is undoubtedly due, in part, to the general conservatism of the community and to the fact that a certain knowledge of Fourier optics is essential to understand properly the method (both its strengths and limitations). However, the lack of acceptance is also due to the fact that speckle interferometry, as originally proposed [Labeyrie, 1970; Dainty, 1985] does not produce an *image* of the object under observation. This drawback is solved by the technique of bispectral imaging discussed in this paper. Another drawback is the fact that speckle interferometry is limited to fairly bright objects — the faintest objects resolved to date, for example the "triple" quasar PG1115+08 [Foy *et al.*, 1985] and the Pluto/Charon system [Baier & Weigelt, 1987], both having visual magnitudes

---

*Blackett Laboratory, Imperial College, London SW7 2BZ, UK.

on the order of 16: this order of limiting magnitude (perhaps as faint as $m_v \approx 20$) is a fundamental limit [Dainty, 1985] governed by the statistics of atmospheric and photon noise and affects any passive technique of imaging through turbulence using only the light emitted by the object under observation.

In 1977, G. Weigelt [Weigelt, 1977] proposed a method he called "speckle masking" that processed the speckle interferometry short-exposure images in such a way as to produce diffraction-limited *images* of the object. The method was not widely recognised, even within the specialised high angular resolution community, probably because it was not understood and perceived to be rather *ad hoc*. However, in 1983 the speckle masking process was identified by Weigelt & colleagues [Lohmann *et al.*, 1983] to be equivalent to a triple correlation of the short-exposure data, or in Fourier space, the bispectrum of the data. Thus the technique is now usually called triple correlation imaging or bispectral imaging. This identification with the bispectrum occurred at a time when there was a considerable resurgence of interest in the bispectrum and other higher order statistics in signal processing, with the result that the technique was rapidly recognised as *the* method of processing short-exposure speckle data to produce diffraction-limited images. The acceptance of the technique was further enhanced when F. Roddier [Roddier, 1986] identified the bispectrum as a generalisation of the concept of "closure phase" which forms the basis of interferometric imaging in radio astronomy.

Although a number of groups throughout the world have contributed to the theoretical understanding of bispectral imaging, and have implemented the technique on large telescopes, the present status and success of the method is due almost entirely to G. Weigelt and his group and the astronomical results shown later are from his observations.

In this paper, we give a short introduction to the bispectrum in §2, followed by a description of its application in astronomical imaging in §3. Some speculations on future developments, particularly in relation to the technique of adaptive optics, are given in the final section.

2. The bispectrum

The bispectrum is one of a family of higher-order polyspectra of random processes: the most familiar is the second-order spectrum, or power spectrum, which, for a statistically stationary random process $i(x)$, is defined as

$$I^{(2)}(u) = \lim \ X \to \infty \ \left\langle \left| \frac{1}{2X} \int_{-X}^{+X} i(x) \ \exp(-2\pi i u x) \ dx \right|^2 \right\rangle . \tag{1}$$

For a stationary random process, the various spectra have to be defined as the limit of $X \to \infty$ since the Fourier transform does not strictly exist. However, in our case, we shall be dealing with (spatially) non-stationary random processes, the short-exposure speckle images, for which a Fourier transform can be defined, and

160

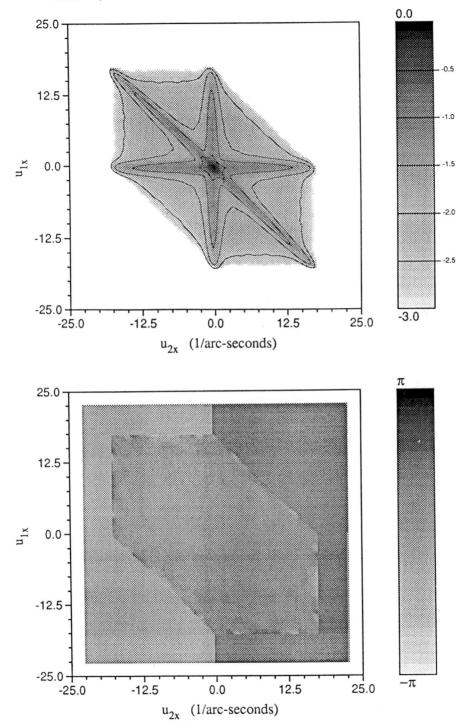

Figure 1. Computer simulations of the modulus (top) and phase (bottom) of the bispectral transfer function [Ayers *et al.*, 1988]. The simulation is for a 2-m telescope, $r_0 \approx 20$ cm and $\lambda = 0.55 \ \mu$m; 10^4 realisations were used.

in this case the power spectrum[1] can be simply defined as

$$I^{(2)}(u) = \left\langle |I(u)|^2 \right\rangle \tag{2}$$

where

$$I(u) = \int_{-\infty}^{+\infty} i(x) \exp(-2\pi i u x) \, dx \tag{3}$$

is the Fourier transform of $i(x)$. The power spectrum does not contain information on the Fourier phase of the signal (at least in any obvious form). The autocorrelation function is the inverse Fourier transform of the power spectrum, and for a non-stationary statistical signal is therefore given by

$$i^{(2)}(x_1) = \left\langle \int_{-\infty}^{+\infty} i(x) i(x + x_1) \, dx \right\rangle. \tag{4}$$

The triple correlation and bispectrum are also Fourier pairs and are defined, for spatially non-stationary images, by

$$i^{(3)}(x_1, x_2) = \left\langle \int_{-\infty}^{+\infty} \int_{-\infty}^{+\infty} i^*(x) i(x + x_1) i(x + x_2) \, dx_1 dx_2 \right\rangle \tag{5}$$

and

$$I^{(3)}(u_1, u_2) = \left\langle I(u_1) I(u_2) I^*(u_1 + u_2) \right\rangle. \tag{6}$$

The key characteristic of the bispectrum is that Fourier phase information about the original signal is retained and it is this feature that makes it useful in a variety of applications in signal processing. For example, in the analysis of acoustic signals, a time-skew stationary signal (one that sounds differently when it is played backwards) has different bispectra for the forward and reverse versions, whereas the power spectra of the two are identical. References [Nikias & Raghuveer, 1987] and [Lohmann & Wirnitzer, 1984] provide an introduction to the bispectrum in signal processing and optics.

One obvious problem in dealing with the bispectrum is the fact that the bispectrum of a one-dimensional signal is two-dimensional, and that of a two-dimensional signal (e.g. an image) is four-dimensional. The bispectrum of a real non-stationary process has a 12-fold symmetry and this reduces the storage requirement: nevertheless, a considerable computer memory is required to store the complete bispectrum of an image ($\approx$ 5 Gwords for a single 512×512 image).

3. Bispectral imaging through turbulence

3.1. Theory

In conventional speckle interferometry, the average power spectrum of the image is equal to the product of the power spectrum of the object and the "speckle transfer function" so in one-dimensional notation we may write

$$I^{(2)}(u) = O^{(2)}(u) \times T^{(2)}(u) \tag{7}$$

---

[1]More correctly, the *energy* spectrum

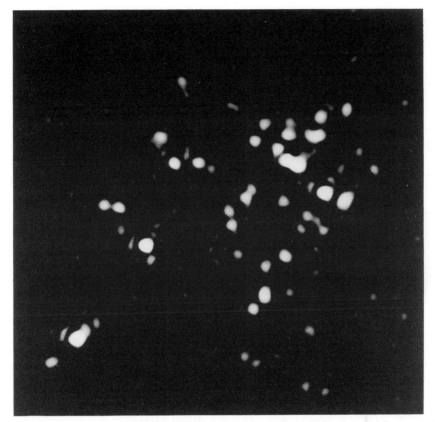

Figure 2. Bispectral image of R136 [Pehleman *et al.*, 1991]; the field covered is 4.9 by 4.9 arcsec and the faintest star resolved is $V_{710} \approx 16.4$ mag.

In this (and similar equations below), it is understood that the object spectrum is not statistical and therefore the averaging operation is unnecessary. It can be shown [Korff, 1973; Dainty, 1985] that $T^{(2)}(u)$ is non-zero to the diffraction-limit of the telescope (even if the telescope optics has small aberrations) and thus the power spectrum of the object can be found provided that a reliable estimate of the speckle transfer function can be made using a reference star: this is not completely straightforward because the atmospheric statistics are non-stationary in time and in angular position on the sky.

Given a measurement of the object power spectrum $|O(u)|^2$, it is not possible to reconstruct a map of the object in a straightforward way because the Fourier phase is unknown. The problem of estimating the object from its Fourier modulus is called the "phase problem" and is encountered in many other fields of science. In one dimension, there is no unique solution to the phase problem (except in certain special cases). However, in two dimensions, solutions that are effectively unique, and

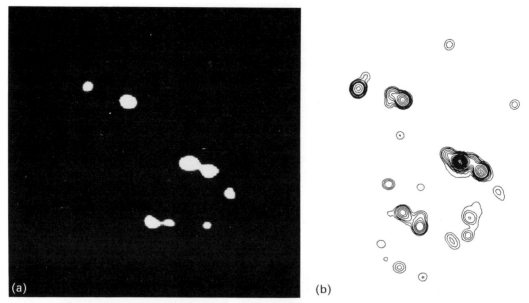

Figure 3. Comparison of bispectral image of R136a [Weigelt & Baier, 1985] with HST image [Weigelt *et al.*, 1991].

computable, can exist. This subject as applied to astronomical optical interferometry is reviewed in references [Bates & McDonnell, 1986] and [Dainty & Fienup, 1985] and some recent advances in algorithms are given in [Lane, 1991].

Finding the bispectrum of the image intensity circumvents the phase problem and allows the Fourier phase to be estimated, albeit somewhat indirectly. The bispectrum of the image is related to that of the object by a relation analogous to Eqn (7):

$$I^{(3)}(u_1, u_2) = O^{(3)}(u_1, u_2) \times T^{(3)}(u_1, u_2) \tag{8}$$

where $T^{(3)}(u_1, u_2)$ is the bispectral transfer function. It can be shown [Wirnitzer, 1985] that, to a good approximation, the phase of the bispectral transfer function is zero; this means that the phase $\beta(u_1, u_2)$ of the bispectrum of the image *equals* that of the object bispectrum. From the definition of the bispectrum, the phase of the bispectrum is related to the phase $\alpha(u)$ of the object spectrum by

$$\beta(u_1, u_2) = \alpha(u_1) + \alpha(u_2) - \alpha(u_1 + u_2). \tag{9}$$

Note that Eqn (9) is identical to a "closure phase" [Jennison, 1958] used in radio astronomy. Originally, a recursive algorithm [Lohmann *et al.*, 1983; Bartelt *et al.*, 1984] was used to recover the Fourier phase $\alpha(u)$ of the object intensity from

164

the measured bispectral phase $\beta(u_1, u_2)$. It is more satisfactory to use some kind of least squares procedure to estimate $\alpha(u)$ from the measurements of $\beta(u_1, u_2)$ [Meng et al., 1990], [Glindemann et al., 1992]. The modulus of the Fourier phase can be estimated from the power spectrum in the normal way.

In fact, only a fraction of bispectrum is of practical use for estimation of the Fourier phase of the object. Fig. 1 shows the result of a computer simulation estimate [Ayers et al., 1988] of the modulus and phase of the bispectral transfer function (the calculation was two-dimensional but only a single sub-plane is shown). As stated above, the phase of this function is zero. The modulus of the bispectral transfer function, however, only has large values along narrow bands close to the line of symmetry. The fraction of the bispectrum that is useful depends on D/r_0 (D is the telescope diameter and r_0 is the Fried parameter [Fried, 1966]) and on the level of noise.

The axes of symmetry in the bispectrum are almost equal to the power spectrum and it follows, after a rigorous analysis [Ayers et al., 1988], that the signal-to-noise ratio in the regions close to the axes (i.e. the useful part) is essentially equal to the signal-to-noise ratio of the power spectrum. The significance of this result is the following: if the data are good enough to give a interpretable power spectrum by conventional speckle interferometry, then they are probably good enough to yield a useful *image* of the object by bispectral analysis. This is a rather vague statement and further work is required to understand how noise in the measured bispectrum influences the final reconstructed image.

3.2.   Results

Since the first publication on speckle interferometry in 1970 [Labeyrie, 1970], there have been approximately 250 papers reporting astrophysical results using single-aperture interferometry [Foy, 1991; Gezari et al., 1990]. Bispectral imaging techniques have only been used more recently, some of the more noteworthy results being on η-Carinae [Weigelt & Ebersberger, 1986], Pluto/Charon [Baier & Weigelt, 1987] and R136 [Weigelt & Baier, 1985; Weigelt et al., 1991].

Fig. 2 gives an idea of the present "state of the art" of bispectral imaging. It is a 4.9 by 4.9 arcsec diffraction-limited image of R136 [Pehleman et al., 1991], the central object of the 30 Doradus nebula in the LMC obtained from ≈ 6000 frames taken on the 1.54-m Danish/ESO telescope at La Silla. A total of 46 stellar components have been identified, the faintest of which is $V_{710} \approx 16.4$ mag; the dynamic range is $\approx 30{:}1$. This is the first ground-based optical image (by *any* technique) that begins to approach the remarkable images produced routinely by aperture synthesis in radio astronomy. Figs. 3(a) and (b) show the central part of this field, R136a, taken from Earth using bispectral imaging and using the Hubble Space Telescope [Weigelt et al., 1991]: all of the components first observed by Baier and Weigelt [Weigelt & Baier, 1985] in 1985 using bispectral imaging are confirmed by the HST picture.

165

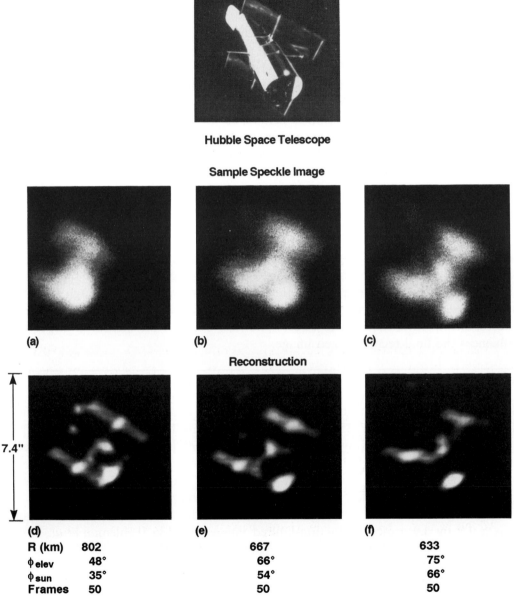

Figure 4. Bispectral images of the Hubble Space Telescope [Lawrence *et al.*, 1991].

It is believed that bispectral imaging has also been used extensively for the observation of man-made satellites. Fig. 4 shows one example [Lawrence *et al.*, 1991] — diffraction-limited images of the Hubble Space Telescope taken with a 1.6-m telescope using only 50 frames for each image.

4. Future developments

In examining possible future developments of bispectral imaging, one has to consider the whole range of high angular resolution imaging techniques currently being developed for use on large telescopes; these include image sharpening, adaptive optics and non-redundant array imaging, as well as bispectral imaging.

The trade-off between non-redundant aperture imaging on a large telescope and filled aperture bispectral imaging is now fairly well understood. Non-redundant aperture imaging has a well-behaved transfer function close to unity at all spatial frequencies and has a better signal-to-noise ratio for bright objects. Bispectral imaging appears to be better for fainter objects. These two techniques are complementary and perhaps some hybrid version will emerge in the future.

At the present time, the technique of adaptive optics has caught the imagination of astronomers, due in large part to the published results of the COME-ON project [Rousset et al., 1990]. Because both r_0 and the timescale of the turbulence are larger at longer wavelengths (both are proportional to $\lambda^{6/5}$), adaptive optics systems are fundamentally easier to implement in the infra-red. At visible wavelengths, the technique will only give a *partial* correction, at least if the guide star used for the wavefront sensing is an astronomical object. This immediately opens up two possibilities: either bispectral imaging could be used together with adaptive optics or bispectral imaging could be supplemented by wavefront data in off-line processing.

This latter technique, called "speckle deconvolution" [Primot et al., 1990], is particularly attractive at light levels that are too low for closed-loop operation of adaptive optics. Using a dichroic mirror with narrow-band ($\approx 10 - 20$ nm) transmission for the speckle data, most of the light could be used by the wavefront sensor. A 15th magnitude object gives a flux at the telescope pupil on the order of 10^{-3} photon/s/Å/cm², so that in white-light flux (say $\Delta\lambda = 1000$Å) one has on the order of 1 photon per r_0^2 per 10 ms (compared to an image-plane flux on the order of 100 photons per 100Å per 10 ms) for a 4-m telescope containing 10^3 r_0^2 areas. How to combine the image-plane and pupil-plane data optimally at these flux levels is still an open question.

Adaptive optics techniques at visible wavelengths are severely limited by the photon flux of the guide star. However, if laser guide stars [Foy & Labeyrie, 1985; Murphy et al., 1991] become technically feasible, then adaptive optics will be capable of diffraction-limited imaging in the visible provided that a natural reference object brighter than $m_v \approx 19$ mag is within the isoplanatic patch (about 2% sky coverage), this limit being imposed by the flux necessary to correct tilt fluctuations not correctable with a laser guide star. Passive imaging techniques will still have a role, particularly the bispectrum which is insensitive to wavefront tilt and therefore may not require the reference object. Indeed, it seems certain that post-detection processing will always be required in order to extract the maximum amount of information from actively-corrected images.

167

Acknowledgements

We wish to thank Professor G. Weigelt for providing Figs. 2 and 3, and Dr T. Lawrence for Fig. 4.

A personal note

I wish to express my deep gratitude to Professor Charles Wynne, to whom this book is dedicated, for introducing me to the subject of speckle interferometry in 1971. He drew my attention to A. Labeyrie's original discovery and urged me to follow it up. I was a little slow on the uptake but Professor Wynne kept reminding me of the technique almost every time we met in the corridor at Imperial College (i.e. daily) and eventually I saw the light. I am very grateful to him not only for kindling my interest in this field, but also for his unfailing personal support over the past two decades. *(Chris Dainty)*

References

[Ayers et al., 1988] Ayers G.R., Northcott M.J., Dainty J.C., *Knox-Thompson and triple correlation imaging through turbulence. J. Opt. Soc. Amer. A, Vol. 5 (1988), p. 963.*

[Baier & Weigelt, 1987] Baier G., Weigelt G., *Speckle interferometric observations of Pluto and its moon Charon on seven different nights. Astron. Astrophys., Vol. 174 (1987), p. 295.*

[Bartelt et al., 1984] Bartelt H., Lohmann A.W., Weigelt G., *Phase and amplitude recovery from bispectra. Appl. Optics, Vol. 23 (1984), p. 3121.*

[Bates & McDonnell, 1986] Bates R.H.T., McDonnell M.J., *Image Restoration and Reconstruction, Oxford University Press, 1986.*

[Dainty, 1985] Dainty J.C., *Stellar speckle interferometry. in Laser Speckle and Related Phenomena, Ed. Dainty J.C., Springer-Verlag, 2nd Edition, 1985.*

[Dainty & Fienup, 1985] Dainty J.C., Fienup J.R., *Phase retrieval and image reconstruction for astronomy in Image Recovery: Theory and Application, Academic Press, 1985.*

[Foy, 1991] Foy R., *Single aperture interferometry. Proc. International Conference on High Angular Resolution Interferometry, ESO, Garching, 1991.*

[Foy & Labeyrie, 1985] Foy R., Labeyrie A., *Feasibility of adaptive telescope with laser probe. Astron. Astrophys., Vol. 152 (1985), p. L29.*

[Foy et al., 1985] Foy R., Bonneau D., Blazit A., *The multiple QSO PG1115+08: a fifth component? Astron. Astrophys., Vol. 149 (1985), p. L13.*

[Fried, 1966] Fried D.L., *Optical resolution through a randomly inhomogeneous medium for very long and very short exposures.* J. Opt. Soc. Amer., Vol. 56 (1966), p. 1372.

[Fugate *et al.*, 1991] Fugate R.Q. et al., *Measurement of atmospheric wavefront distortion using scattered light from a laser guide star.* Nature, Vol. 353 (1991), p. 144.

[Gezari *et al.*, 1990] Gezari D.Y., Roddier F., Roddier C., *Spatial Interferometry in Astronomy.* NASA Reference Publication 1245, (1990).

[Glindemann *et al.*, 1992] Glindemann A., Lane R.G., Dainty J.C., *Estimation of binary star parameters by model fitting the bispectrum phase.* J. Opt. Soc. Amer. A, Vol. 9 (1992), p. 543.

[Haniff, 1991] Haniff C.A., *Least-squares Fourier phase estimation from the modulo 2π bispectrum phase.* J. Opt. Soc. Amer. A, Vol. 8 (1991), p. 134.

[Humphreys *et al.*, 1991] Humphreys R.A. et al., *Atmospheric turbulence measurements using a synthetic beacon in the mesospheric sodium layer.* Opt. Lett., Vol. 16 (1991), p. 1367.

[Jennison, 1958] Jennison R.C., *A phase sensitive interferometer technique for the measurement of the Fourier transform of the spatial brightness distribution of small angular extent.* Mon. Not. R. Astr. Soc., Vol. 118 (1958), p. 276.

[Korff, 1973] Korff D., *Analysis of a method for obtaining near-diffraction-limited information in the presence of atmospheric turbulence.* J. Opt. Soc. Amer, Vol. 63 (1973), p. 971.

[Labeyrie, 1970] Labeyrie A., *Attainment of diffraction-limited resolution in large telescopes by Fourier analysing speckle patterns in star images.* Astron. Astrophys., Vol. 6 (1970), p. 85.

[Lane, 1991] Lane R., *Phase retrieval using conjugate gradient minimisation.* J. Mod. Optics, Vol. 38 (1991), p. 1797.

[Lawrence *et al.*, 1991] Lawrence T. et al., *Proc. International Conference on High Angular Resolution Interferometry*, ESO, Garching, 1991.

[Lohmann *et al.*, 1983] Lohmann A.W., Weigelt G., Wirnitzer B., *Speckle masking in astronomy: triple correlation theory and its applications.* Appl. Optics, Vol. 22 (1983), p. 4028.

[Lohmann & Wirnitzer, 1984] Lohmann A.W., Wirnitzer B., *Triple correlations.* Proc. IEEE, Vol. 72 (1984), p. 899.

169

[Matson, 1991] Matson C.L., *Weighted least-squares phase reconstruction from the bispectrum. J. Opt. Soc. Amer. A, Vol. 8 (1991), p. 1905.*

[Meng *et al.*, 1990] Meng J., Aitken G.J.M., Hege E.K., Morgan J.S., *Triple correlation sub-plane reconstruction of photon-address stellar images. J. Opt. Soc. Amer. A, Vol. 7 (1990), p. 1243.*

[Murphy *et al.*, 1991] Murphy D.V. *et al., Experimental demonstration of atmospheric compensation using multiple synthetic beacons. Opt. Lett., Vol. 16 (1991), p. 1797.*

[Neri & Grewing, 1988] Neri R., Grewing M., *AIT-MCP-Speckle camera observations of the multiple star cluster R136a. Astron. Astrophys., Vol. 196 (1988), p. 338.*

[Nikias & Raghuveer, 1987] Nikias C.L., Raghuveer M.R., *Bispectrum estimation: a digital signal processing framework. Proc. IEEE, Vol. 75 (1987), p. 869.*

[Pehleman *et al.*, 1991] Pehleman E., Hofmann K.-H., Weigelt G., *Proc. International Conference on High Angular Resolution Interferometry, ESO, Garching, 1991.*

[Primmerman *et al.*, 1991] Primmerman C.A. *et al., Compensation of atmospheric optical distortion using a synthetic beacon. Nature, Vol. 353 (1991), p. 141.*

[Primot *et al.*, 1990] Primot J., Rousset G., Fontanella J.C., *Deconvolution from wavefront sensing: a new technique for compensating turbulence-degraded images. J. Opt. Soc. Amer. A, Vol. 7 (1990), p. 1958.*

[Roddier, 1986] Roddier F., *Triple correlation as a phase closure technique. Opt. Comm., Vol. 60 (1986), p. 350.*

[Rousset *et al.*, 1990] Rousset G. *et al., First diffraction-limited astronomical images with adaptive optics. Astron. Astrophys., Vol. 230 (1990), p. L29.*

[Weigelt, 1977] Weigelt G., *Modified astronomical speckle interferometry, speckle masking. Opt. Comm., Vol. 21 (1977), p. 55.*

[Weigelt & Baier, 1985] Weigelt G., Baier G., *R136a in the 30 Doradus nebula resolved by holographic speckle interferometry. Astron. Astrophys., Vol. 150 (1985), p. L18.*

[Weigelt & Ebersberger, 1986] Weigelt G., Ebersberger J., *Eta Carinae resolved by speckle interferometry. Astron. Astrophys., Vol. 163 (1986), p. L5.*

[Weigelt *et al.*, 1991] Weigelt G. *et al., First results from the faint object camera: high resolution observations of the central object R136 in the 30 Doradus nebula. Astrophys. J., Vol. 378 (1991), p. L21.*

[Welch & Gardner, 1991] Welch B.M., Gardner C.S., *Effects of turbulence induced anisoplanatism on the imaging performance of adaptive astronomical telescopes using laser guide stars. J. Opt. Soc. Amer. A, Vol. 8 (1991), p. 69.*

[Wirnitzer, 1985] Wirnitzer B., *Bispectral analysis at low light levels and astronomical speckle masking. J. Opt. Soc. Amer. A, Vol. 2 (1985), p. 14.*

14
COAST – a progress report

Peter Warner *

Abstract

A short description of COAST, the Cambridge Optical Aperture Synthesis Telescope, is presented, including design principles, current status, and plans for further development.

1. Introduction

The Cambridge Optical Aperture Synthesis Telescope (COAST) will soon be an array of four small telescopes operating as Michelson interferometers on baselines up to 100 m; using closure phase techniques it will produce images in the visible and near-infrared with a resolution of up to 1 milliarcsec. The single baseline (2.7 m) operating at the moment has been producing stellar fringes regularly since June 1991.

The principles behind the design of COAST have been demonstrated by the production of diffraction-limited images from the 4.2-m William Herschel Telescope. The method of using non-redundant masking to measure closure phases and some striking pictures of changing features of the surface of Betelgeuse have been presented by Buscher *et al.* [1990] and Wilson [1992].

2. Design

The design of COAST can be broken down into a number of key components. The novel arrangement of the telescopes can be seen in Fig. 1 and consists of a 50-cm siderostat feeding a fixed horizontal 40-cm f/5.5 Cassegrain reducing telescope. The resulting parallel light beams are reflected *via* the central telescope into the optics laboratory. Here, in order to see any fringes, the light paths from the star *via* each telescope must be equalised. This is achieved by a path-compensation system consisting of roof mirror mounted on a trolley whose position (and velocity) is servo-controlled to about 1μm. Further details of this system and the laser interferometer that is used to control it can be found in Boysen & Rogers [1992].

---

*Mullard Radio Astronomy Observatory, Cavendish Laboratory, Madingley Road, Cambridge CB3 0HE, UK.

Figure 1. The Cambridge Optical Aperture Synthesis Telescope – COAST.

A further prerequisite for the detection of high contrast fringes is the superposition of the images from each telescope. The blue light from each telescope is used to measure the apparent position of the star. Errors are then corrected using the small piezo-actuated mirror at the back of each telescope. Currently the light from each telescope is focused onto different patches of a single CCD which is then read every 12 ms. As shown in Fig. 2 this fast autoguiding servo reduces image motion at frequencies down to 10 Hz by about an order of magnitude. This same CCD is

COAST Auto−guider Performance

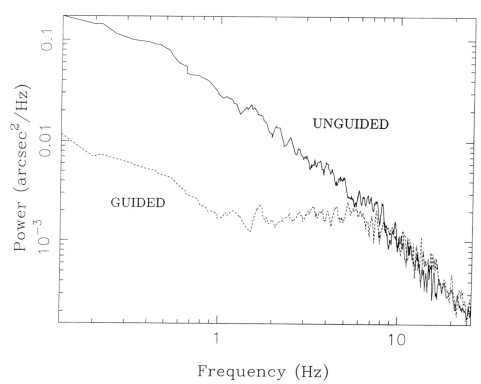

Figure 2. Power spectra of guided and unguided stellar motion.

also used as an acquisition system. A more complete description appears in Cox [1992].

The remaining red light is used to form fringes. With just two telescopes this is easily done with a 50:50 beam-splitting cube in the parallel beam. If the instrument is correctly aligned this pupil is either dark or light and so can be measured by a pair of single element detectors. For this we are using avalanche photodiodes (APDs) which we developed for this purpose [Nightingale, 1991]. By cooling them thermoelectrically to $-35°C$ it is possible to operate in a photon counting mode with a DQE of 40% at 800 nm.

The layout of the optics within the COAST lab is shown in Fig. 3. A crucial step in the operation of the telescope and the detection measurement of stellar fringes is its internal alignment. This is done by setting the siderostats normal to the axis of the fixed Cassegrain optics so that light from an artificial star is reflected back through the instrument to the detectors. With the autoguiders running it is possible to use the path compensation trolley to move the white light fringes through the

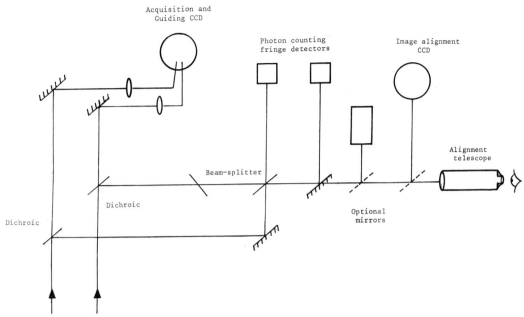

Figure 3. The layout of the acquisition and guiding, fringe detection and alignment optics used for the two COAST telescopes.

APDs. When the internal paths are matched to within the coherence of the system, fringes are detected. In practice we connect the output of the APD directly to a loudspeaker and fringes can be heard easily above the photon noise. An example is shown in Fig. 4.

The original detection of stellar fringes from Vega was done in essentially the same way. The path compensation trolley was moved to a position slightly in advance of that required for the star's hour angle and declination; the rotation of the earth then sweeps the fringes past the detector at frequencies up to 200 Hz. The short (0.2 s) burst of fringes is readily heard. They can be seen in Fig. 5. Further examples can be found in Baldwin [1992].

3. Development

The next vital step in the development of COAST will be the installation of the third telescope and the second path compensation trolley. Once they are operating successfully and fringes have been found between the third telescope and one of the others, we will start trying to measure closure phase or the triple product. This also requires a different and more general arrangement for beam-combining; the layout we propose uses four beamsplitting plates at 10-deg incidence and is shown in Fig. 6. This simple arrangement uses all the light but to measure the complex visibility of each of the six different baselines we get from the four telescopes we

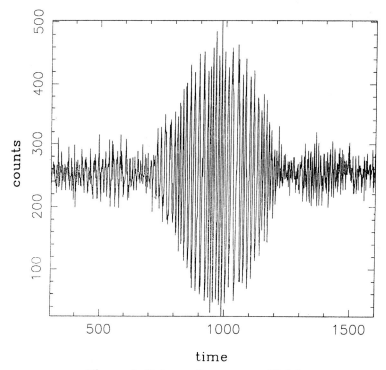

Figure 4. Fringes from an artificial star.

must "label" the light from each before combination. We do this by modulating its phase in a unique way and then using phase-sensitive detection on the outputs. We will use a sawtooth modulation of the path or phase which is introduced by the path-compensation system: its amplitude is only a few wavelengths and its frequency 160 Hz. With three telescopes there is just a single closure phase to measure, but with four telescopes there are three distinct closure phases. In these measurements the unwanted contributions from the atmosphere and the telescope cancel out leaving a phase which tells us something about the object itself. They are then processed in exactly the same way that VLBI data is to produce diffraction limited images.

The layout of COAST will eventually consist of many sets of telescope foundations in the form of a **Y**. One telescope will sit at the apex and the others will be moved along the arms to distances of up to 100 m. This gives reasonable aperture-plane coverage without requiring observations at low elevations.

We look forward to using this complete instrument during 1993 to image a wide range of astronomical objects. A resolution of about 1 milliarcsec should be attainable on stars and binary systems as faint as 10 mag in the visible or near infrared.

fringes A Boo 27/07 2

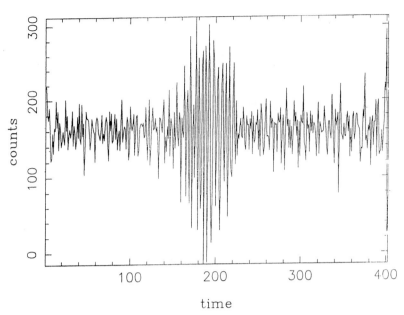

Figure 5. Fringes from Arcturus at 830 nm with a bandwidth of 40 nm.

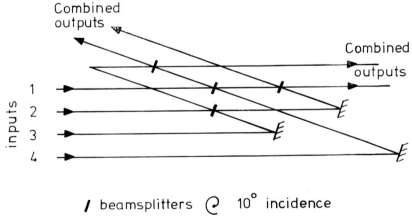

Figure 6. The arrangement of beam combiners to be used for four COAST telescopes.

Acknowledgements

The COAST team currently consists of John Baldwin, Roger Boysen, Graham Cox, Craig Mackay, John Rogers, Peter Tuthill, Peter Warner, Donald Wilson and Richard Wilson.

References

[Baldwin, 1992] Baldwin J.E., *Proc. ESO Conference on High Resolution Imaging by Interferometry II, ESO, 1992.*

[Boysen & Rogers, 1992] Boysen R.C., Rogers J., *Proc. ESO Conference on High Resolution Imaging by Interferometry II, ESO, 1992.*

[Buscher *et al.*, 1990] Buscher D.F., Haniff C.A., Baldwin J.E., Warner P.J., *Mon. Not. R. astr. Soc., Vol. 245 (1990), p. 7P.*

[Cox, 1992] Cox G., *Proc. ESO Conference on High Resolution Imaging by Interferometry II, ESO, 1992.*

[Nightingale, 1991] Nightingale N., *A new silicon avalanche photodiode photon counting detector module for astronomy. Experimental Astronomy, Vol. 1 (1991), p. 407.*

[Wilson, 1992] Wilson R.W., *Proc. ESO Conference on High Resolution Imaging by Interferometry II, ESO, 1992.*

15
Adaptive Optics – The MARTINI System

John Major [*]

Abstract

An outline is given of atmospheric seeing and the limits it sets to imaging by telescopes. This leads on to the principles and details of design of the MARTINI[1] adaptive optics system. It is followed by a presentation of its main technical and astronomical results.

1. Introduction

In the optical path between stellar light arriving at the top of the atmosphere and finally imaging in the focal plane of an astronomical telescope the wavefront becomes distorted from three main sources. Turbulence in the atmosphere above the telescope introduces local variations to the refractive index such that neighbouring rays of light travel slightly different optical paths. The result is that on arrival at the telescope the wavefront is generally inclined to its original plane and moreover is quite corrugated. Both the tilting of the wavefront and the degree of corrugation change on time-scales of 10–20 ms. The consequences of these are that (from wavefront tilting) the image moves about in the focal plane rapidly and that the instantaneous image is broadened (from the corrugations) to the extent that it may break up into many speckles. Together these lead to images which are broad compared to the diffraction limit of the aperture of the telescope. This is termed atmospheric 'seeing'. Added to the affects of the atmosphere are similar distortions and movements of the image arising from convection currents from within the dome of the observatory and from vibrations and movements of the telescope and its structures. There is some dispute over the relative importance of these extra contributions to image distortion but there is no doubt that inadequate control of the disposal of heat within the dome can have significant, deleterious effects.

Over the last 20 years or so there has been an increasing interest in understanding the physics behind this image deterioration and, following this understanding, more experimental interest in adaptive optics, concerned with sensing these movements and broadening of stellar images and attempting to correct them, optically, in real-time.

[*]Department of Physics, University of Durham, South Road, Durham DH1 3LE, UK.
[1]MARTINI – Multi-Aperture Real-Time Image Normalisation Instrument.

179

1.1. Atmospheric seeing – general background

In selecting mountain sites for astronomical observatories attention must be given to the shape of the mountain and to the prevailing winds. Good observing sites such as Mauna Kea and La Palma are old volcanic islands approximating to inverted cones above relatively flat seas. If the air-flows over these sites were laminar then air velocity, temperature, density and refractive index would be relatively constant at a given height. However, turbulence does occur and this leads to a mixing of the levels of the air-flow resulting in variations of index, for example, over the scale of the turbulence. Within the turbulence, conditions are such that smaller sized turbulent eddies are generated which vary the index on these smaller scales. Smaller and smaller eddies are generated until the Reynold's Number falls below the critical value for instability. The smallest eddies are finally dissipated through friction. The large eddies are formed at the outer scale of turbulence, L_0, and they are finally dissipated at the inner scale of turbulence, l_0. There is some controversy over the size of the outer scale of turbulence. Some years ago it was thought that L_0 was of the order of 100–1000 m but recent measurements and analysis suggest that it is of the order of 1 to 10 m. The inner scale of turbulence is the order of mm or cm.

As a stellar wavefront passes down through the atmosphere it experiences both the vertical and the horizontal variations in refractive index which have been induced by the turbulence. On arrival at the observatory the wavefront will have become corrugated on spatial scales corresponding to the range between the inner and outer scales of turbulence. Further, over the size of the aperture the wavefront may be locally tilted. It is customary to think in terms of the Taylor hypothesis in which the distorted wavefront is regarded as temporarily 'frozen' within the airstream but as the airstream moves across the aperture of the telescope the spatial variations of the distorted wavefront give rise to temporal variations. In this case the distortions will vary on time-scales related to the windspeed.

Dimensional analysis [Tatarski, 1961] shows that the structure function for refractive index in the turbulent atmosphere (that is, the mean square variation of index between all points distance r apart in the atmosphere) is given by:

$$D_n(r) = \left\langle |n(r_1) - n(r_2)|^2 \right\rangle = C_n^2 r^{2/3}$$

As the wavefront passes through the atmosphere these index variations give rise to phase variations along the wavefront. These depend upon the column integral of the index of the atmosphere above the wavefront. There is a corresponding structure function for the phase variations:

$$\begin{aligned} D_\phi(r) &= \left\langle |\phi(r_1) - \phi(r_2)|^2 \right\rangle = C_\phi^2 r^{5/3} \\ &= 6.88 \left(\frac{r}{r_0} \right)^{5/3} \end{aligned}$$

where, finally, the phase variations have been normalised through r_0. The coherence parameter r_0 is defined at a wavelength, λ_0, of 0.5 μm where it is typically in the range of 10 to 20 cm. It is the size of the circular area over which the mean square phase variations are just over 1 radian squared (that is this circular field is flat to about 1/6 of a wavelength). For an extended wavefront there is incoherence on scales larger than r_0 with the result that the image that is formed has a size of $\theta_0 = \lambda_0 / r_0$ which is about 1 arcsec when r_0 is 10 cm.

As stellar light passes through the atmosphere its dispersion delays the red wavefront relative to the blue so that it arrives a little later. However, despite the temporal delay the spatial variations, measured in microns, along the wavefronts are essentially identical and must be of the order of a few microns at most. This has been termed polychromaticity. Consequently the mean square variation of phase in the red wavefront will be scaled down relative to that in the blue because of its longer wavelength. This can be expressed by a coherence length, r_λ which is correspondingly larger. From the expression for the phase structure function above it is straightforward to show that:

$$r_\lambda = \left(\frac{\lambda}{\lambda_0}\right)^{6/5} r_0$$

For example if r_0 is 10 cm then the coherence length at 2.2 μm is approximately 60 cm. The image size that it will produce is given by:

$$\theta_\lambda = \frac{\lambda}{r_\lambda} = \frac{\lambda_0}{r_0} \left(\frac{\lambda_0}{\lambda}\right)^{6/5}$$

with the result that the image in the red is slightly smaller than in the blue.

The time-scale of the variations of the wavefront phases will increase with wavelength in the same way as the coherence length. So at 2.2 μm, for example, the coherence length and the timescale of seeing are expected to be about six times larger than at 0.5 μm. This is why progress on astronomical full-wave front correction has, so far, only been possible in the infra-red at wavelengths of about 1–3 μm.

However, the motion of the image in the focal plane is independent of wavelength; its rms value is related to the overall image size λ_0 / r_0 at l_0 and is given approximately by:

$$\alpha = 0.42 \frac{\lambda_0}{r_0} \left(\frac{r_0}{D}\right)^{1/6}$$

where D is the diameter of the telescope aperture. Since the motion is independent of wavelength then so must be its timescale.

So a picture emerges of an image, which is instantaneously broadened or speckled, moving about in the focal plane. The broadening or speckling is worse in the blue and it changes on a shorter time-scale than in the red. However, the overall movements of the image (and the timescales) in the focal plane are independent of wavelength. The time integrated image is the combined effect of the instantaneous broadening being further broadened by the movements in the focal plane.

A much more detailed account of atmospheric seeing is given in Roddier [1981].

181

1.2. The physics of stellar imaging

An appropriate way of examining the imaging of stars by a ground-based telescope under the canopy of the Earth's atmosphere is through the optical transfer functions of the various optical components through which stellar light passes on its way to the focal plane. It is convenient to regard the very distant star as a δ-function spike of intensity. The Fourier transform of this is a spectrum of spatial frequencies stretching uniformly over all possible values. If these are superposed, in phase, then the original spike will be reconstituted.

However, the spatial frequency response of a telescope aperture (its modular transfer function) falls off with spatial frequency so that there is no transfer of frequencies beyond that expected from the Abbé theory of image formation, that is beyond a maximum spatial frequency of D/λ where D is the aperture diameter and λ is the wavelength. When the stellar wavefront has been truncated by the aperture through which it passes it can no longer contain spatial frequencies beyond the maximum value. Consequently when an image is reconstituted by superimposing the remaining spectrum of frequencies it can no longer correspond to the original spike; instead the familiar diffraction pattern is formed.

The atmosphere, through its distorting affects on the wavefront, only transfers frequencies over a relatively low range. The modulation transfer function of the atmosphere has been expressed by Fried [1978] as:

$$\text{MTF}(\omega) = \exp\left[-3.44\left(\frac{\lambda\omega}{r_\lambda}\right)^{5/3}\right] \quad \text{and since } r_\lambda = \left(\frac{\lambda}{\lambda_0}\right)^{6/5} r_0$$

$$= \exp\left[-3.44\left(\frac{\lambda_0}{\lambda}\right)^2\left(\frac{\lambda_0\omega}{r_0}\right)^{5/3}\right]$$

which gives a better image at long λ

It is seen that spatial frequencies up to about r_λ/λ only are transferred through the atmosphere. When this very reduced spectrum, devoid of high spatial frequencies, is superposed an image is produced which is very broad compared to the diffraction limited image.

The role of adaptive optics is, somehow, to restore the component of high spatial frequencies so that on imaging fine detail is reproduced.

A large part of the power of the distortions of the wavefront is in the rapid movement of the image arising from the tilt of the wavefront. Its removal is the important first step in image correction and MARTINI sets out to do this. Experimentally the succeeding images of a star are superimposed in the focal plane by sensing where the instantaneous image lies and tip-tilting a small mirror in the optical beam line to move the image to its nominal focal spot. Fried [1978] showed

182

that the corresponding modulation transfer function is:

$$\mathrm{MTF}(\omega) = \exp\left[-3.44\left(\frac{\lambda_0}{\lambda}\right)^2\left(\frac{\lambda_0\omega}{r_0}\right)^{5/3}\left[1-\left(\frac{\lambda\omega}{D}\right)^{1/3}\right]\right]$$

The extra factor in the exponent restores some of the high spatial frequencies and particularly at the cutoff frequency of the aperture, D/λ. Physically this can be understood by remembering that $\mathrm{MTF}(\omega)$ is also the mean value of the cosine of the difference in phase between points separated by $\lambda\omega$ over the aperture. Removing wavefront tilt makes the mean cosine equal to unity for points separated by a diameter and so $\mathrm{MTF}(D/\lambda)$ will become unity.

Now the overall MTF is the product of the modulation transfer functions of all the contributing optical components. The main contributions come from the aperture of the telescope and from the atmosphere. The image to which these will give rise is found by taking the Fourier transform of the overall MTF. This has been done in Fig. 1 [Dunlop, 1987; Doel, 1990] which shows the FWHM of the expected image. The characteristic feature is that there is an optimum aperture at which the image size (after correction for image motion, that is wavefront tilt) is expected to be about three times smaller than the uncorrected image size. The actual gain that is possible depends upon the outer scale of turbulence. When this is small the wavefront tilting becomes reduced and there is less improvement available. In Fig. 1 the outer scale of turbulence that has been assumed is very large.

The optimum aperture has a size of four times the coherence length. In the visible region, where r_0 is typically 10–20 cm in size the optimum aperture is 40–80 cm in diameter. At 2.2 μm where r_λ is about six times larger the optimum diameter becomes 2.4–4.8 m. If correction is attempted at 2.2 μm (for a typical mirror aperture) then it is sufficient to remove the tilt over the whole aperture. However, for the visible spectrum, where the optimum is quite small, it is necessary to correct the tilt over a multiplicity of sub-apertures each of which is of the optimum size. This is the basis of MARTINI which is designed to give wavefront tilt correction for the 4.2-m William Herschel Telescope in the visible spectrum. In the prototype version there are six sub-apertures. If a single tip-tilt mirror, corresponding to the whole of the WHT aperture is used in the visible then the improvement in image size is, from Fig. 1, only about 20%.

By working with $4r_0$ sub-apertures MARTINI is able to correct for image motion with guide stars of about 13 mag.

2. The development of MARTINI as a semi-common user device

2.1. The adaptive system

2.1.1. General system

The optical hardware of MARTINI now consists of two breadboards on which the optical elements are permanently mounted and pre-aligned. The two parts are

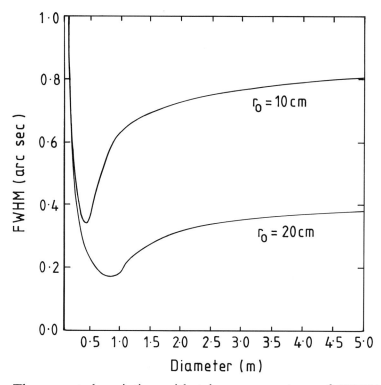

Figure 1. The expected variation with telescope aperture of FWHM for images corrected for their motions. There is an optimum size of aperture at $4r_0$.

separately transported to WHT where they are kinematically locked together at the optical table in the GHRIL (Ground-based High Resolution Imaging Laboratory) which is located on one of the Nasmyth platforms. Fig. 2 shows an outline of its main optical parts.

The stellar field (containing a guide star) is imaged at the Nasmyth focus where 1 arcsec corresponds to $230\,\mu m$. This field is re-imaged, with unit magnification, to a plane 1.2 m downstream on the GHRIL optical table by reflection first at the adaptive mirrors and secondly at the toroid mirror. In front of the six adaptive mirrors is a wheel containing a series of masks each defining a set of six sub-apertures (corresponding to 0.2 to 1.4 m) which can be chosen to be of optimum size by matching to the prevailing seeing. At the re-imaged Nasmyth focus there are six independent images moving according to their associated wavefront tilts. As they move through each other, interference fringes are instantaneously seen on the EEV camera which examines the guide star. At the re-imaged focus a perforated beam-splitter allows the light from the six images of the guide star to be transported and focused on an IPD (Imaging Photon Detector) whereas the rest of the field is reimaged on the prime CCD detector. The motions of the six guide star images are

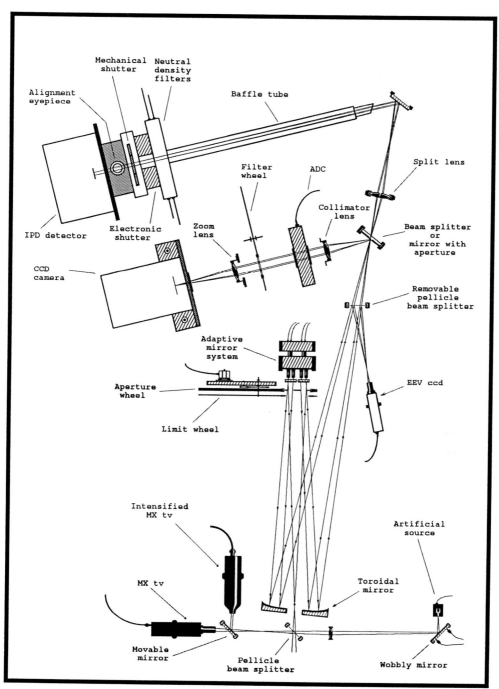

Figure 2. The present optical layout of MARTINI 2.

sensed at the IPD and these are servo-looped back to the piezo-crystals which drive the adaptive mirrors, inclining them, to reduce the motions of the images of the guide star. As the motion of the guide star is corrected then so is the motion of the rest of the stellar field. Consequently at the CCD focus the seeing motions of the six separate, but superimposed, stellar fields are reduced and the images sharpened.

As each photon is recorded at the IPD the controlling 68020 chip quickly decides whether it is a signal or noise photon by checking whether it has fallen within an acceptable distance of a sub-image. If it has then an algorithm will estimate the best position that the mirror for that sub-aperture should be moved to to intercept the next photon. The chip takes about $70 \mu s$ to handle an x or y coordinate and one chip serves all six sub-images. It begins to saturate at a total rate of about 7 kHz which is in excess of the rate expected for a 13 mag guide star. When high rates are required (for example in examining the detailed seeing motions from which the power spectrum is determined) then the magnification of the system is increased by a further factor of four to the IPD which then defines a 3 arcsec field. Then only one sub-image is recorded and the total rate is given over to this single image.

Currently MARTINI has a maximum field of view of 104 arcsec which is set by the 25 mm aperture in the toroid mirror. In recent astronomy runs the field at the CCD camera has been about 40 arcsec in size which corresponds roughly to the isoplanatic patch of the device. Over the isoplanatic patch guiding adaptively on a star of 13 mag leads to sharpening of the star-field by a factor of about two.

2.1.2. Detailed aspects of the optics

The toroid mirror: Much of the special optics have been produced in-house. The toroid mirror has been deliberately made to introduce negative astigmatism in order to counteract the positive astigmatism that is produced because the toroid mirror is used off-axis. It was prepared in-house by Dr David Brown by stressing a spherical mirror about a diameter so that it gave good imaging in the off-axis position. It was then repolished to a sphere at a local optical shop after which the stressing was removed and it relaxed to a toroid. After rotating through a right-angle the toroid is correctly positioned.

The split lens: The separation of the guide star images on the IPD is brought about by use of a 'split' lens (see Fig. 2). This was also produced in-house by mounting a 20 mm diameter, 100 mm focal-length doublet lens in a jig, then precisely sawing it into six sectors and remounting the sectors so that they touched again. Each sector matches a mask of an adaptive mirror and the width of the sawcuts is such that the images of the wavefronts defined by the masks are separated onto the corners of a hexagon of about 10 mm side on the 40 mm diameter cathode of the IPD. There is a magnification of about 13 times onto the IPD whose 1024 pixels across its diameter define a field about 12 arcsec in size. An important benefit of this final stage of magnification is that the image is made large enough so that the chance of successive photons falling into the same microchannel of the IPD is small during the 15 ms recovery time of the activated microchannel. The IPD is protected both by mechanical and electronic shutters. Further to these there is a set of neutral density filters just in front of the IPD, used with brighter guide stars to reduce the photon intensity to about 1 kHz per sub-aperture which is the rate corresponding to a 13 mag star. The IPD is cooled with circulating antifreeze at about $-25°C$ so that the noise is reduced to about 60 counts/s over the whole of the IPD field.

Beam splitters: The beam-splitter at the reimaged Nasmyth focus was made in-house by aluminising a thin glass plate but for a central circular patch 1 mm (4 arcsec) or 1.5 mm (6 arcsec) in diameter. The aperture allows 96% of the guide star light to pass to the IPD and 4% to the CCD whereas all the light of the general field, apart from the central patch, is reflected to the CCD. The 'black hole' corresponding to this can be seen on the CCD frames shown later with the faint (4% of the light) guide star at its centre.

All parts are motorised wherever possible so that remote adjustments and changes can be made and run from the WHT control room by the Unix workstation.

2.1.3. Conjugate plane for the adaptive mirrors

The position of the adaptive mirrors (0.6 m downstream from Nasmyth focus) is chosen to be conjugate to seeing at 3.8 km above the WHT. The wavefronts of light from a guide star and from a field star which are equally distorted by the seeing at that height will be exactly superimposed in the conjugate plane. Consequently when a mirror is adjusted to correct the tilt of the wavefront of the guide star it will tilt that of the field star by exactly the same amount and so alter both wavefronts equally. If the seeing is above or below 3.8 km then the wavefronts of the guide and field stars overlap but are not precisely superimposed in the plane of the adaptive mirrors. However, for the compromise position that has been chosen the wavefronts will overlap by at least 80% and so correcting the guide star will still give good correction for the field star.

With the conjugate at 0.6 m beyond Nasmyth focus there is a demagnification of about 70 times from the seeing layer. Consequently the whole aperture is covered by a system of adaptive mirrors which are 60 mm in diameter and an optimum Fried aperture of 40 cm size is replaced by a mask of 6 mm diameter in front of the adaptive mirror. Similarly if the wavefront at the 4.2 m aperture is inclined at 1 arcsec then the adaptive mirrors will need to be tilted by 70 arcsec. Since the individual adaptive mirrors have a maximum mask aperture of about 20 mm then the mirror can be tilted by 70 arcsec with a $7 \mu m$ extension of the piezo-crystals. The piezo-crystals have a maximum extension of $17 \mu m$ at 150 volts.

2.1.4. Other important features of the system

CCD and collimator: From the re-imaged Nasmyth the beam splitter reflects the stellar field to the CCD where it is re-imaged by use of a collimator. The collimator

187

allows the introduction of filters without upsetting the focusing and the introduction of the Imperial-College-designed automatic dispersion corrector. The rear lens of the collimator is a zoom lens which allows the plate scale of the CCD to be changed without refocusing. Indeed, in image sharpening it is critical to focus the CCD as precisely as possible. If the CCD is 1 mm beyond focus in the $f/11$ beam then about 0.5 arcsec of defocus is introduced. The focusing is described later. Our own CCD has been mounted in a cradle which can be moved in x,y,z under micrometer control to assist the precise positioning.

The artificial star: This is a small but bright, battery driven, LED which with subsidiary optics produces an $f/11$ image at Nasmyth focus. It is used to help with the general alignment of the total optics, in making the adaptive mirrors parallel prior to running, in focusing both the CCD and the secondary mirror of WHT and, through the extra agency of a 'wobbling mirror', to produce image motions of the artificial star for testing correction algorithms. For this latter purpose the wobbling mirror is driven by a stream of real image centroids recorded on an earlier observing run at WHT. This is realistic both in size and in the power spectrum of the motions of the artificial images.

The adaptive mirrors must be made parallel prior to running adaptively. The images from the six sub-apertures of the artificial star are viewed on the EEV camera. In general they are close together but not overlapping. Through the computer keyboard the piezo crystals are first run through a large number of hysteresis cycles of diminishing amplitude. Then the six images are individually moved until they overlap properly; this is signalled by the production of a characteristic hexagonal interference pattern. The piezo voltages for this are recorded so that subsequent to running adaptively the mirrors can be repositioned. In this adjustment to parallel the positions of the six sub-images on the IPD are recorded. These become the nominal zeros to which the seeing-shifted images are corrected during adaptive running. Because the positions on the IPD are determined without the use of the primary and secondary mirrors, then correction of real star images to these positions will remove any static shifts brought about by misalignment of the telescope mirrors.

An important use of the artificial star is in focusing. First, by direct observation the artificial star is adjusted to lie in the centre of the aperture of the toroid mirror. Its image on the CCD is recorded and by adjusting the x and y micrometers of the CCD cradle the image is centred there. To focus the CCD in the z-direction all but two of the six masks in front of the adaptive mirrors are covered leaving a two-aperture Hartmann–Shack screen. Each of these masks leads to images of the artificial star which superimpose exactly at focus. On either side of focus the separation of the images increases linearly with the defocus. The focusing is carried out semi-automatically. The cradle micrometers are motorised and, together with the making of CCD exposures at a series of z-positions and the analysis of their image positions, is under computer control. The cradle is driven to the computer-calculated position of focus. In this position the image size of the stationary artificial star is better than 0.2 arcsec.

The secondary mirror is positioned in a similar way but using a real star. The CCD is left in its focused position and the secondary mirror is adjusted in z, again recording on the CCD the two out-of-focus images. Analysis of the separations of these images gives the focus position of the secondary mirror. In this position the images of the real and the artificial stars coincide at Nasmyth focus.

Acquisition: Via initial beam splitters, the 5 arcmin field at Nasmyth can be examined on TV or intensified TV cameras. It is used initially at WHT to find the several stars required to set up the pointing model for the telescope at Nasmyth where there is no guidance system. The artificial star image is used here to define the centre of the field.

Although the pointing model used for tracking is good there is still slippage of the image over periods of an hour. With MARTINI this is automatically corrected by the adaptive mirrors slewing to take up the slippage. Indeed by measuring the slew the adaptive system could be used to provide a guidance system for WHT Nasmyth.

Automatic Fried curve: One of the aperture masks consists of six sub-apertures with sizes running from the equivalent of 0.2 m to 1.4 m. Each sub-aperture is backed by a neutral density filter so that the light flux from each is approximately the same. By using this mask and recording IPD data a Fried curve can be produced very quickly (in about 1 min). The minimum image size that this displays determines the optimum sub-aperture at which the system should be run.

2.2. System control

Whereas the earlier version of MARTINI required four PCs within the GHRIL control room to control its separate functions, these are now handled by a single Unix-based workstation in the WHT control room. Accompanying this are a real-time display of the six sub-images of MARTINI and a display of the Nasmyth field for acquisition.

The control software uses the network-transparent X-Windows system and can be used from the GHRIL control room (during set-up) or from the WHT control room (during operation) without hardware changes and with an identical interface for the user.

3. Results from MARTINI

MARTINI can be run either (or simultaneously) as a real-time or post-exposure sharpening device. For the latter the x, y positions of all the photons are recorded for subsequent off-line data analysis. This latter mode ('photon dumping') allows many of the characteristics of seeing to be examined. Some typical results from post-exposure sharpening are described below and are followed by real-time sharpening results. Detailed accounts of all these will be found in Brown *et al.* [1988], Clegg *et al.* [1991], Doel *et al.* [1989, 1990, 1991a, 1991b], Shanks *et al.* [1990], and Tanvir *et al.* [1991].

3.1.   Post-exposure analysis

3.1.1.   Fried curve

The principle on which MARTINI has been designed is that there is an optimum sub-aperture for which tip-tilting is most effective. Fried showed that this would occur at $4r_0$ where the correction for tilt would result in images being about three times sharper. Unfortunately this optimum is only 0.4 m to 0.6 m in size. Hence in MARTINI several sub-apertures are used simultaneously to increase the fill of the primary aperture.

In Fig. 3 the sizes of images before and after sharpening are shown as a function of aperture size. Clearly there is a minimum in image size occurring at about 0.5 m aperture. Compared to the uncorrected images there is an improvement by a factor of about two in the sharpness. The fitted curve drawn through the corrected images is calculated from Fried's 'short exposure' MTF of the atmosphere where r_0 is a parameter of the fit. Here r_0 is about 0.15 m and for this value the uncorrected image size is calculated using Fried's 'long exposure' MTF of the atmosphere. However the calculation is sensitive to the so-called outer scale of turbulence, L_0 (whereas the corrected images are not). When this is used as a parameter of the fitting then a value of the order of 2 m to 5 m is required. This supports the finding of J. Vernin (who used quite different methods) that the outer scale is less than 10 m [private communication]. However, it should be pointed out that aberrations in the telescope system might be present which prevent the Fried curve from reaching its true minimum. The effect is to shift the apparent optimum aperture to a smaller value and correspondingly to a smaller value of r_0. This would lead to a reduced apparent value of L_0.

3.1.2.   Post-exposure image sharpening

Several examples of post-exposure sharpening are shown in the Fig. 4. They show the sharpening of binary stars. The best performance has been post-exposure sharpening of a single star from FWHM of 0.7 arcsec to 0.24 arcsec.

3.1.3.   Power spectra of image motion

These encapsulate the characteristics of seeing and define what properties an adaptive system must have. The temporal variations in the positions of the centroids of moving (but uncorrected) images are determined from the recorded photons at the IPD. From the Fourier transform of their autocorrelation the power spectral density

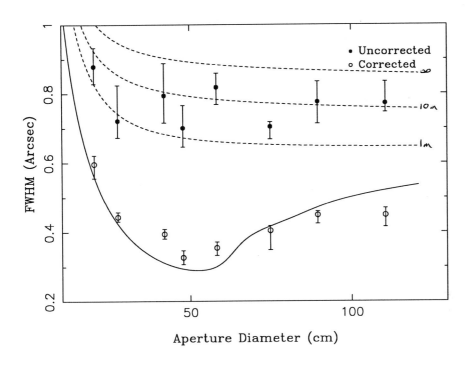

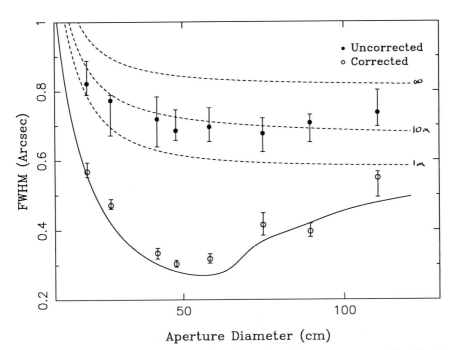

Figure 3. Fried curves compared with theoretical long exposure fits for L_0 values of ∞, 10 m and 1 m.

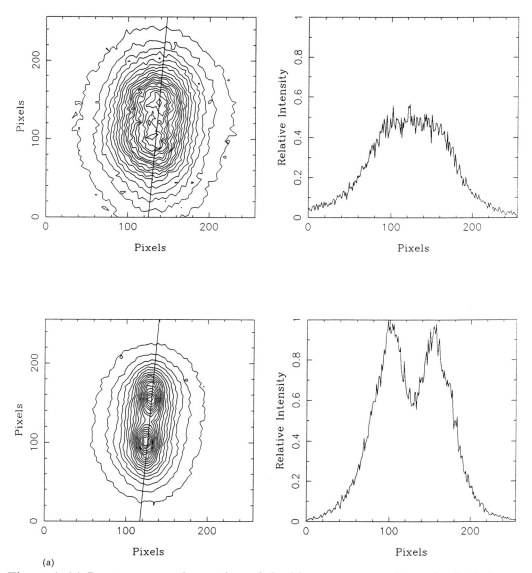

(a)

Figure 4. (a) Post-exposure sharpening of the binary star Zeta-Boo; the field shown is 2.6 arcsec.

is determined. This gives the power (in arcsecs squared) of the motion as a function of the frequencies which are present in the motion. This tells what frequencies of motion need to be removed to produce an acceptable sharpened image. In Fig. 5, examples are shown. In Fig. 5a the typical Kolmogorov spectrum is demonstrated with an initial $f^{-2/3}$ dependence falling away at $f^{-8/3}$ beyond a knee whose position is related to the wind speed blowing in the turbulent layer. In Fig. 5b a more usual spectrum is seen falling away as $f^{-5/3}$. According to Martin (1987) this arises when

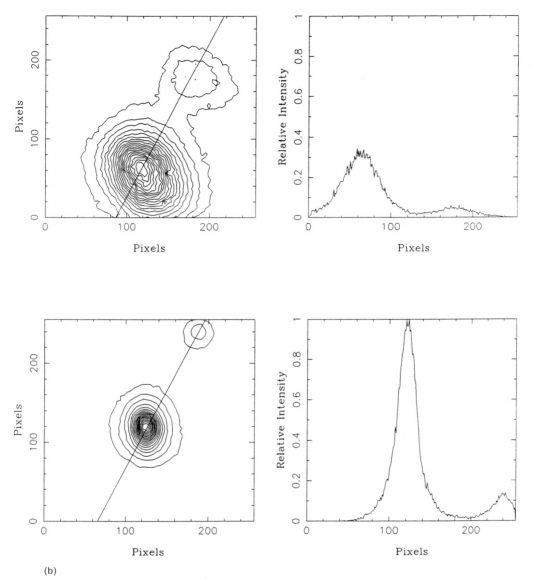

Figure 4. (b) Post-exposure sharpening of the binary star 25-CVN; the field shown is 3.4 arcsec.

the seeing is distributed between several turbulent layers within which there are different wind speeds.

In both spectra most of the power is concentrated at frequencies below about 30 Hz. Consequently this is the frequency range with which the adaptive mirrors and control algorithms must cope.

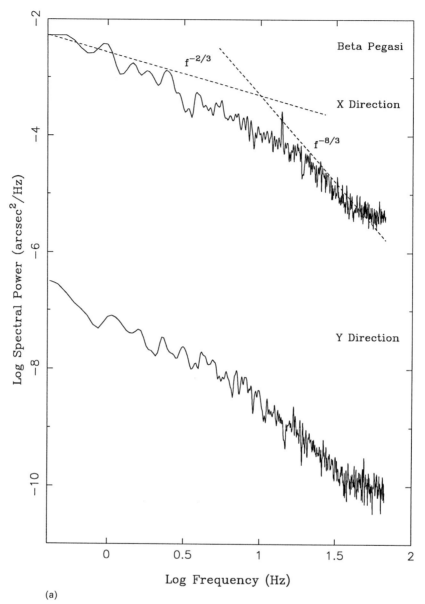

Figure 5. (a) The x and y power spectra of image motion for Beta Pegusi. The y power spectrum is shifted by a factor of 10^{-4} for clarity.

3.2. Real time analysis

3.2.1. Power spectra of motion

To be effective the adaptive mirrors must remove temporal frequencies of motion below about 30 Hz. In Fig. 6 a recent example of this is shown in which the guide

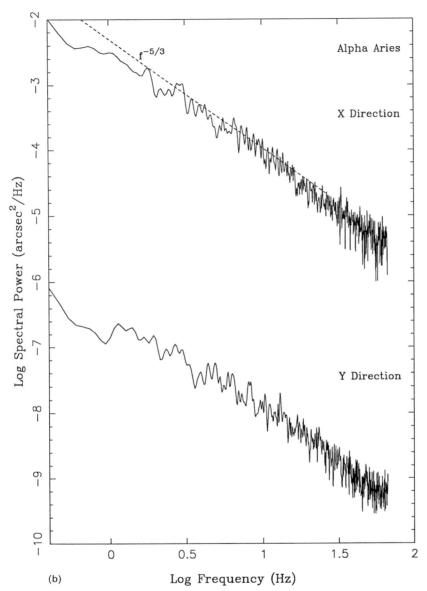

Figure 5. (b) The *x* and *y* power spectra of image motion for Alpha Aries. The *y* power spectrum is shifted by a factor of 10^{-4} for clarity.

star was corrected with photon rates of about 1700 and 7000 photons per sec per subaperture. Clearly the low frequencies are removed when compared to the results for the same, but uncorrected guide star image. At the higher photon rate the removal is more effective since the centroids are being determined more precisely. This would argue towards high photon rates. However, there is not a *pro rata* return

195

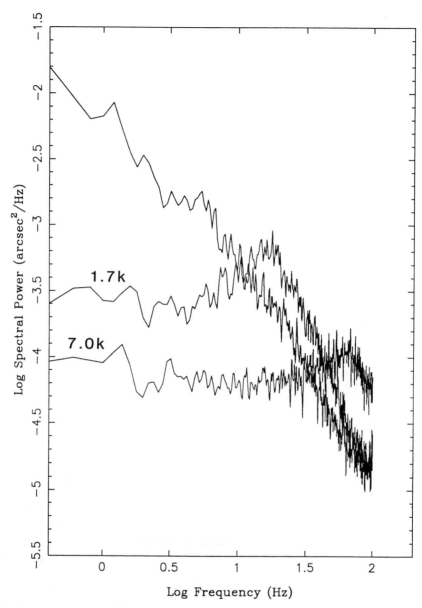

Figure 6. Power spectra of image motion for Alpha-Boo before and after real-time image correction on the IPD wavefront sensor for various photon rates.

for high photon rates for although higher frequencies are removed their contribution to image size becomes progressively weaker. However, there is a more important effect which is seen in the figure that needs to be addressed. This effect is the injection of noise above the cross-over frequency. Simulations show that this is due

196

to the algorithm employed in the adaptive control. To some extent this is responsible for real-time sharpening being slightly less effective than post-exposure sharpening.

3.2.2.   Real-time image sharpening

Similar examples to those given earlier are shown in Figs. 7 and 8 though the degree of sharpening is slightly reduced.

3.2.3.   Isoplanatic patch size

The angular range over which the corrections to a guide star are appropriate to the other stars in a field is the isoplanatic patch. Some star fields of two and three stars have been examined and show that the isoplanatic patch is in excess of 30 arcsec. An example where there is still good correlation after 9 arcsec is shown in Fig. 8. This would imply an isoplanatic patch of at least twice this size. Indeed work by others suggests that this might be as large as a few arcmin for tilt distortion.

4.   Common user astronomy results

There have been two collaborative runs using MARTINI for astronomy with colleagues within this Department and with colleagues at the RGO. For the present purpose only the visual impact of the sharpened sources is presented. Unfortunately when the system has been used for astronomy rather than for the more technical characteristics of seeing then the prevailing conditions have not been very good. Whereas MARTINI is designed to sharpen from 1 arcsec downwards it has been necessary to run it when the seeing has been about 2 arcsec. This is a severe test of the system and particularly of the robustness of the controlling algorithm. The sharpening by a factor of two is still observed but it is now from about 2 arcsec to 1 arcsec. A more potent effect is that the optimum aperture has to be considerably reduced. To work at acceptable photon rates as before a brighter guide star is needed. In the sharpening that follows the guide stars were still 13 mag and consequently the apertures used were larger than the optimum value. Despite that the value of an image sharpening device is clearly demonstrated.

4.1.   M92 cluster

This is shown before and after sharpening in Fig. 9. In the centre of the picture is the shadow of the beam splitter aperture with the guide star at its centre. By examining common features on the pair of pictures the effect of sharpening by a factor of two is clearly to be seen.

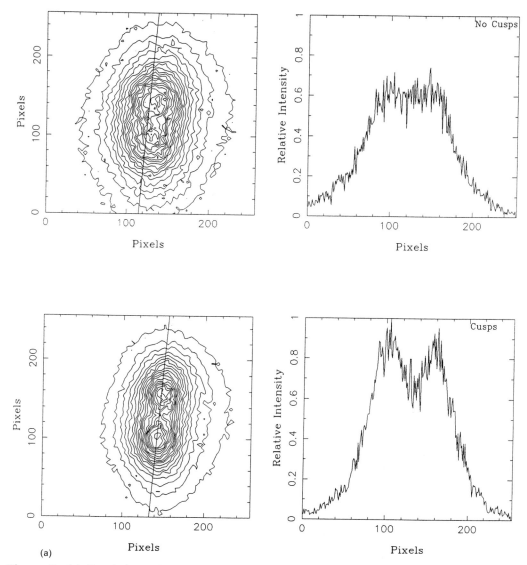

Figure 7. (a) Real-time sharpened image of the binary Zeta-Boo. The binary itself was used for the guide star.

4.2. The Virgo cluster

This is shown only in its sharpened form in Fig. 10. Initial results on Virgo are given in Shanks *et al.* [1992]. The size of the universe is still generally considered to be uncertain to about a factor of two. By observing individual stars in the Virgo cluster galaxies much better estimates of its distance can be found. Since the relative

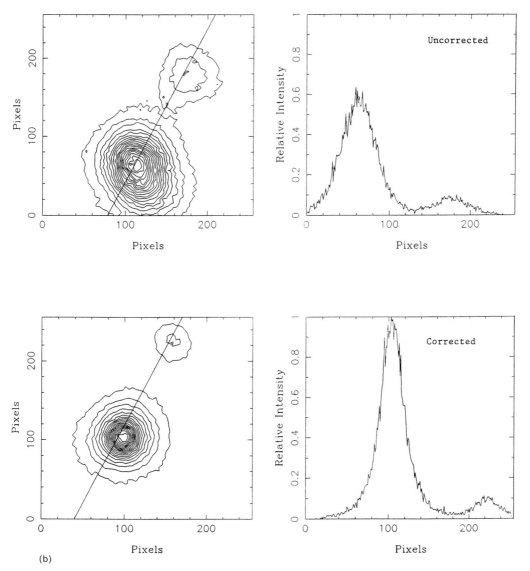

Figure 7. (b) Real-time sharpened image of the binary 25-CVN. The binary itself was used for the guide star.

distance between Virgo and more remote clusters of galaxies is well known, this would, in turn, lead to more precise estimates of distances right across the universe.

In order to see individual stars in Virgo galaxies, images with higher resolution than commonly available from ground-based telescopes are required. Using MARTINI at WHT in June 1991 the Virgo cluster NGC4523 has been observed. Although observing conditions were good for only about 30 min, one of the highest-

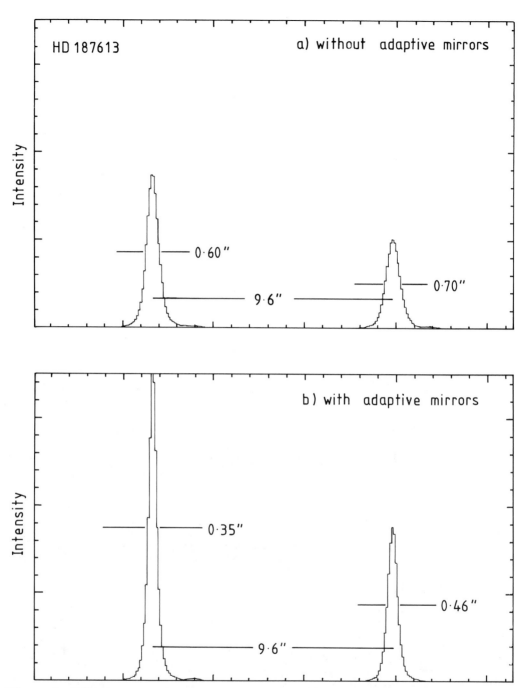

Figure 8. CCD images of HD187613 recorded with MARTINI with and without adaptive correction. The two stars are 9.6 arcsec apart and the brighter was used as the guide star.

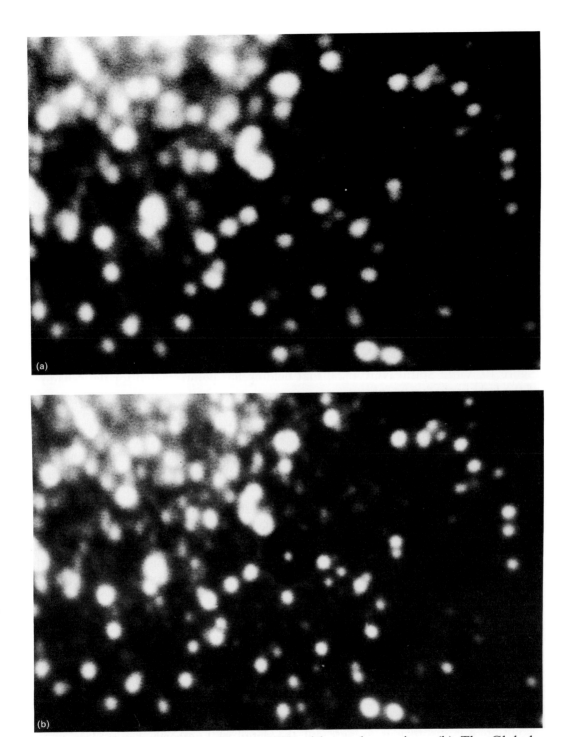

Figure 9. (a) The Globular Cluster M92 without sharpening. (b) The Globular Cluster M92 with real-time sharpening.

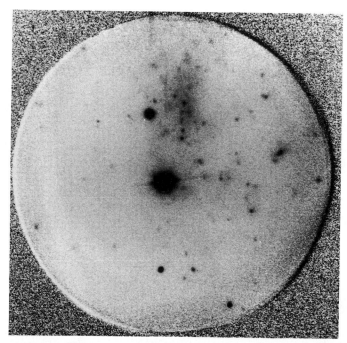

Figure 10. Virgo Spiral NGC4523. Flat fielding has compensated for the "shadow" of the hole in the beam splitter.

resolution images so far, of this galaxy, was obtained. The brighter objects in the image are foreground stars in our galaxy. Several faint objects are candidate stars stars in NGC4523 and a provisional analysis suggests that they imply a surprisingly low distance of only 13–15 Mpc.

If the result is confirmed by further observations with longer integration times, then it could have profound implications for cosmology. This arises because the size of the universe, inferred from this result, would, assuming the Big-Bang, give an age of the universe of only around 7 Gyr. This would be in glaring contradiction to the measured ages of the galaxy's oldest stars which are thought to be 16 Gyr old.

4.3. Abell 78

This 'born again' planetary nebula is shown in Fig. 11 taken in [OIII] light. It resolves, for the first time, the hydrogen-poor inner region into a series of radial filaments or 'jets' containing knots. Again the guide star can be seen inside its surrounding black field. In previous pictures of this object [Jacoby, 1979] the resolution was insufficient to pick out the detail of knots which are close to the guide star.

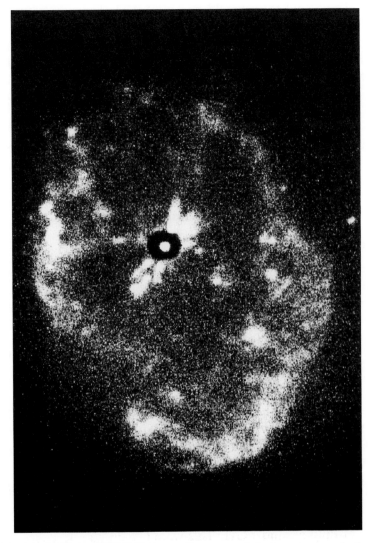

Figure 11. Abell 78 in [OIII] light; a 'born again' planetary nebula; see text.

5. Development of MARTINI

The next stages of development of MARTINI will be along two main directions. First it is intended to improve the response of the system by changing from a controlling 68020 processor to a 68040. This should give a factor 10 in computing speed. This is required so that more sophisticated algorithms can be incorporated in the system. Secondly, it is intended to try to phase the 6, $4r_0$ sub-apertures. This should yield images with a central core less than 0.1 arcsec in size but surrounded by side bands. The Strehl ratio would be about 0.27. There are other, though less fundamental, hardware changes that will be made.

A serious limitation of MARTINI is that with six sub-apertures 10–20% only of the WHT aperture is covered. So a necessary development would be to replace the six sub-aperture system by a 68-segment mirror. With this extension, many of the problems which will be need to be solved in a fully-correcting system will be tackled.

The declassification of results from the SDI programme in the USA are confirming that full-wave correction at all wavelength scales will be possible. It is very important for the UK to capitalise on the release of this information and move quickly to wavefront correction over the whole aperture. This will probably be a mixed system, initially, with near-diffraction correction in the infra-red and image sharpening of the MARTINI kind in the visible.

Acknowledgements

It is a pleasure to acknowledge the support of this work by SERC through research grants and research studentships, the use of the William Herschel Telescope of the Royal Greenwich Observatory at the Spanish *Observatorio del Roque de los Muchachos* of the *Instituto Astrofísica de Canarias* and the help of our colleagues in the Electronics and Mechanical Workshops and in the Microprocessor Centre at Durham.

References

[Brown *et al.*, 1988] Brown D.S., Doel A.P., Dunlop C.N., Major J.V., Myers R.M., Purvis A., Thompson M.T., *Image Stabilisation – the Martini Project. ESO Conference on Very Large Telescopes and their Instrumentation, Garching, 1988.*

[Clegg *et al.*, 1991] Clegg R.E.S., Devaney M.N., Doel A.P., Dunlop C.N., Major J.V., Myers R.M., Sharples R.M., 1991, *Astron. Astrophys., (1991), in press.*

[Doel, 1990] Doel A.P., *PhD Thesis, University of Durham, UK, 1990.*

[Doel *et al.*, 1989] Doel A.P., Dunlop C.N., Major J.V., Myers R.M., Purvis A., Thompson M.T., *Gemini (Newsletter of the RGO), No. 26 (1989), p. 20.*

[Doel *et al.*, 1990] Doel A.P., Dunlop C.N., Major J.V., Myers R.M., Purvis A., Thompson, M.T., 1990, *Stellar Image Stabilisation using Piezo-driven Active Mirrors. SPIE Proceedings of the Conference on Advanced Technology Optical Telescopes, Vol. 1236 (1990), No. 19.*

[Doel *et al.*, 1991a] Doel A.P, Dunlop C.N., Major J.V., Myers R.M., Sharples R.M., *MARTINI: System Operation and Astronomical Performance. SPIE Inter-*

national Conference on Active and Adaptive Optical Systems (San Diego), Vol. 1542 (1991a), No. 31.

[Doel et al., 1991b] Doel A.P., Dunlop C.N., Major J.V., Myers R.M., Sharples R.M., *MARTINI: Sensing and Control System Design. SPIE International Conference on Active and Adaptive Optical Systems (San Diego), Vol. 1542 (1991b), No. 46.*

[Dunlop, 1987] Dunlop C.N., *PhD Thesis, University of Durham, UK, 1987.*

[Fried, 1978] Fried D.L., *Proc. IAU Colloq. No. 50, 1978, p. 4.1 - 4.43.*

[Jacoby, 1979] Jacoby G.H., *Publ. Astron. Soc. Pacif., Vol. 91 (1979), p. 754.*

[Martin, 1987] Martin H.M., *Publ. Astron. Soc. Pacif., Vol. 97 (1987), p. 1215.*

[Roddier, 1981] Roddier F., *The effects of atmospheric turbulence in optical astronomy. in Progress in Optics XIX, Ed. Wolf E., 1981.*

[Shanks et al., 1990] Shanks T., Tanvir N.R., Doel A.P., Dunlop C.N., Myers R.M., Major J.V., Redfern R.M., Devaney M.N., O'Kane P., *The distance of the Virgo cluster via image sharpening. 2nd Rencontres de Blois: Physical Cosmology, Eds Blanchard A. et al., 1990.*

[Shanks et al., 1992] Shanks T., Tanvir N.R., Major J.V., Doel A.P., Dunlop C.N., Myers R.M., *High-resolution imaging of Virgo cluster galaxies - II. The brightest stars detected in NGC4523. Mon. Not. R. astr. Soc., (1992), in press.*

[Tanvir et al., 1991] Tanvir N.R., Shanks T., Major J.V., Doel A.P., Dunlop C.N., Myers R.M., Redfern R.M., O'Kane P., Devaney M.N., *High-resolution imaging of Virgo cluster galaxies - I. Observational techniques and first results. Mon. Not. R. astr. Soc., Vol. 253 (1991), p. 21.*

[Tatarski, 1961] Tatarski V.I., *Wave Propagation in a Turbulent Medium, McGraw Hill: New York, 1961.*

16

A Virgo Galaxy Resolved via WHT MARTINI
Image Sharpening

T. Shanks, N.R. Tanvir*, A.P. Doel*, C.N. Dunlop*,
J.V. Major* and R.M. Myers **

Abstract

We have used the MARTINI image sharpening instrument at the William Herschel Telescope to obtain high-resolution images of two Virgo cluster galaxies.
In 0.6 arcsec FWHM seeing we resolved one of these galaxies, NGC4523,
into stars. Assuming the brightest stars are standard candles, the implied distance of this galaxy is 13 ± 2 Mpc. If this corresponds to the Virgo distance
then $H_o \approx 80$ km s^{-1}Mpc^{-1} and for an $\Omega_o=1$, inflationary Universe, the age
of the Universe is uncomfortably short compared to the age of the oldest
stars.

1. Introduction.

Over the past several years our aim has been to resolve Virgo galaxies into stars
to obtain an improved estimate of that cluster's distance [Shanks *et al.*, 1988, 1991;
Tanvir *et al.*, 1991a, 1991b]. As the nearest significant galaxy cluster, Virgo remains
a crucial step on the cosmological distance ladder. Virgo is the closest cluster
containing large numbers of early-type *and* late-type galaxies and so by stepping out
to Virgo from the Local Group using spiral distance indicators, the early-type galaxy
distance scale can be absolutely calibrated. However, the Virgo distance remains a
very uncertain quantity with distance estimates varying in the range 12–22 Mpc. It
was the original aim of the Hubble Space Telescope to detect the brightest stars and
Cepheids in Virgo galaxies. Here we attempt the same project from a ground-based
telescope using the MARTINI image sharpening device.

The feasibility of resolving Virgo cluster galaxies from the ground became obvious
when Cepheid variables were detected in the galaxy M101 in 1.2 arcsec seeing from
KPNO observations by Cook *et al.* [1986]. The Cepheid variables placed M101 at a
distance of 7.2 Mpc. Thus a similarly crowded galaxy in Virgo should be resolvable
in 0.6-arcsec seeing if Virgo lies at r<15 Mpc and in 0.3-arcsec seeing if Virgo

*Physics Department, University of Durham, South Road, Durham DH1 3LE, UK.

lies at r<30 Mpc. Of course, by using image sharpening techniques [Freid, 1966] 0.6-arcsec seeing is now frequently available and 0.3-arcsec seeing is occasionally available from high altitude observatories such as the Roque de los Muchachos on La Palma.

The Virgo cluster is a particularly good target for image sharpening since there are many galaxies with an overlapping foreground Galactic star that is brighter than the B=13.5 mag limit for a MARTINI monitor star. The MARTINI instrument is described by Doel *et al.* [1990], and in this volume by John Major and colleagues.

2. Observations

In February 1990 during MARTINI commissioning we made our first observations of the Virgo galaxy IC3583. This galaxy was chosen from the list of Sandage & Bedke [1985] who rated it an excellent candidate for resolution by the Hubble Space Telescope on the basis of its low surface brightness, and it is also overlapped by a B=13.5 mag star, useful for image sharpening. IC3583 is given by Sandage and Bedke as a B_T=13.9 mag, Sm III galaxy with a recession velocity of 1120 km/s, close to the mean recession velocity of Virgo galaxies. The observations immediately followed those of a double star, which suggested that, on this night, the isoplanatic patch diameter was at least 37 arcsec. A 1.25-hr exposure was made on IC3583 with 6× ≈0.9-m apertures. The final seeing measured on the co-added frame was 0.60 arcsec FWHM, compared to an average measurement of 0.8 arcsec FWHM unsharpened. Unfortunately, the Nasmyth beam had to be divided 70% to the IPD and 30% to the CCD using a conventional beam-splitter so throughput to the CCD was one-third of that intended. The limiting magnitude was therefore only R=22 mag but even so a lower limit of 15 Mpc for this galaxy's distance was obtained because few candidate brightest stars were seen [Shanks *et al.*, 1991; Tanvir *et al.*, 1991a; Tanvir *et al.*, 1991b].

In June 1991 we observed another Virgo galaxy, NGC4523, with a similar MAR-TINI set-up with 3× higher throughput due to an improved optical arrangement. The imaging detector used was a large format EEV CCD with a pixel scale of 0.14 arcsec per pixel and a MARTINI field-of-view of 110 arcsec in diameter. We chose to observe NGC4523 because it has a bright B=12.9 mag monitor star available. NGC4523 is identified as an excellent candidate for resolution by Sandage and Bedke where it is classified as a B_T=13.6 mag SBdIII galaxy, within the 6-deg radius Virgo core and with a heliocentric recession velocity of 262 km/s. Our results were based on a 15 min exposure when the unsharpened seeing was 0.8 arcsec FWHM and the sharpened seeing with 6×0.75-m sub-apertures was 0.6 arcsec FWHM. This exposure had a limit for stellar detections of R=23.2 mag. We later also obtained further B and Hα CCD frames of NGC4523 from Isaac Newton Telescope observations.

The R image from the MARTINI observation of NGC4523 is shown in Fig. 1. It can be immediately seen that we have detected many individual objects above

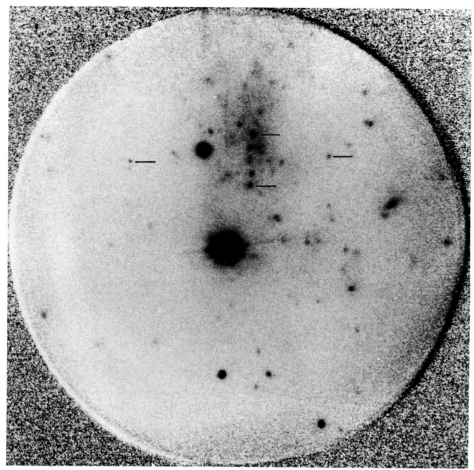

Figure 1. WHT MARTINI R-image of the Virgo galaxy NGC4523. South is up and East is to the left and the diameter of the image is 110 arcsec. The brightest star was used as the monitor for image sharpening. The exposure time was 900 sec and the monitor star seeing is 0.60 arcsec FWHM. The four candidate brightest stars are marked.

the diffuse background light of the galaxy. We carried out semi-automated object detection and found 80 discrete objects with R<23 mag. However, in the magnitude range R<21.5 mag we can reject many objects as being candidate stars on the basis of their non-stellar profile in the high-resolution image. Other compact images show strong Hα and we have classified these as HII regions. After this process four images remained with 20<R<21.5 mag as candidate brightest stars (see Fig. 1). Two of these objects are embedded in the bar of the galaxy. In the central square arcmin where these objects are found we expect 0.5 foreground stars, suggesting that the four stars found in the central region are liable to be associated with the galaxy.

One of the embedded stars has a blue colour of $B - R = 0.2$ mag and R=20.4 mag and we classify this as a candidate blue supergiant; the other three have redder ($B - R = 1.5$ mag) colours and an average magnitude of R $= 21.0$ mag and we classify these as candidate yellow supergiants. In the case of the blue supergiant there remains the possibility that it may be a misidentified blue star cluster or composite system. The redder colours of the candidate yellow supergiants limit the possible effect of any such contamination; also the possibility of confusion with globular clusters seems slight since the specific density of globular clusters seems to be low in LMC-like galaxies such as NGC4523 [Harris, 1991].

Humphreys [1988] has suggested that the brightest yellow supergiants may be good standard candles and these have absolute R (Kron-Cousins) magnitudes in the range −9.6 to −9.7 mag in the LMC [Humphreys & Davidson, 1979]. If this is so then comparing the average magnitude of the three yellow supergiants (R=21.0 mag) to the equivalent quantity in the LMC ($M_R = -9.6$ mag), gives a distance modulus of

$$(m - M)_{AR} = 30.6 \pm 0.3 \quad (13 \pm 2 \text{ Mpc})$$

for NGC4523. Here we are assuming that the errors for yellow supergiant luminosities are similar to those for other classes of supergiant standard candles (e.g. [Sandage & Carlson, 1988]). At this distance the absolute magnitude of the brightest blue supergiant in NGC4523 is $M_R = -10.2$ mag, compared to $M_R = -9.9$ mag for the brightest blue supergiant in the LMC and, given the possible contamination problem, we regard this as in reasonable agreement with the yellow supergiant distance.

3. Discussion

Recently Pierce, McClure & Racine (preprint) have claimed to resolve another Virgo galaxy, NGC4571, in CFHT observations. The brightest stars here have R≈21 mag giving a distance modulus of m−M=30.9±0.2 mag (14.9±1.5 Mpc) in excellent agreement with our result. These results suggest that NGC4571 may lie at approximately the same distance as NGC4523. We now consider the implications for cosmology if the brightest star distances to NGC4523 and NGC4571 prove to be correct. Based on reasonably well established relative distances between Virgo and clusters at higher redshifts (e.g. Coma), Sandage & Tammann [1990] give

$$H_o = 52(21.9 \text{ Mpc}/r_{\text{virgo}}) \text{km s}^{-1} \text{Mpc}^{-1};$$

with $r_{virgo} = 14\pm2$ Mpc (the average of the distances to NGC4523 and NGC 4571) this gives $H_o = 80\pm 10$ km s^{-1}Mpc^{-1}. For the theoretically favoured, $\Omega_o=1$, inflation/CDM model, the age of the Universe is $t_o=8.1\pm1.2$ Gyr; there is then a clear contradiction with the age of globular clusters at 16±2 Gyr (e.g. [vandenBerg, 1988]). Even for low Ω_o, the Universal age is uncomfortably small (12.5 Gyr) compared to the stellar age. Therefore it is hard to understate the implications for cosmology if the Virgo distance is as small as 12–16 Mpc. It is thus of paramount importance to make further observations of NGC4523 and NGC4571, to look for Cepheids and determine their light curves and so firmly establish the distance to these two galaxies. However, even if the brightest star distances to these two galaxies prove correct, the small current size of the Virgo galaxy sample may still allow arguments to be made for lower values of H_o. For example, it should be noted that the two resolved galaxies, NGC4523 and NGC4571, have low velocities with respect to the Virgo mean, with respectively 262 km/s and 342 km/s; the third galaxy, IC3583, which was less well resolved, also happens to have a higher velocity of 1120 km/s. This then provokes the question as to whether NGC4523 and NGC4571 are actually at the mean Virgo distance or whether they might, instead, lie in the foreground.

The possibility of foreground spiral contamination of Virgo has been frequently discussed in the literature, prompted by the very flat spiral velocity distribution between −300 and 2500 km/s in the 6-deg radius cluster core (see Fig. 9 of [Huchra, 1985] where the early and late type velocity distributions are contrasted). The standard dynamical model of Tully & Shaya [1984] suggests that the spiral velocities are highly infall dominated; spirals with $v \approx 2000$ km/s are believed to be at slightly lower distances than the cluster with +1000 km/s infall whereas those at $v \approx 0$ km/s are believed to be at slightly larger distances with −1000 km/s infall. However, it could be asked if other models are possible. In particular, might an asymmetric model be allowed where the above infall explanation of the high-velocity spirals still stood but where the low-velocity spirals were in a foreground cluster(s) sitting on a relatively unperturbed Hubble flow? The usual argument against foreground spiral contamination is simply that if field galaxies are assumed to be distributed around our position isotropically, it then seems unnatural to have an excess of foreground galaxies only in the Virgo line-of-sight. However, this argument is weakened by increasing evidence that the general galaxy distribution may be dominated by filamentary clustering (e.g. [deLapparent et al., 1986]). If Virgo and the Local Group were to be connected by such a spiral-rich filament then severe spiral foreground contamination of Virgo becomes much more plausible. The existence of foreground spiral contamination would obviously also affect the Virgo distance estimate from the Tully–Fisher relation, a distance indicator previously weighted highly because it is calibrated using local spirals with Cepheid distances such as M31. The wide Tully–Fisher relations seen in Virgo by Kraan-Korteweg et al. [1988] and Pierce [1991] can also be taken to argue that the distance distribution of Virgo spirals may be much more complicated than for the early-type galaxies.

Finally, we note that any uncertainty in the Virgo distance also affects the estimate of H_o from the Ursa Major cluster, because Ursa Major only lies at 37 deg from the Virgo line-of-sight. If Virgo is assumed to lie at the the same distance as Ursa Major, the correction to the Ursa Major velocity for the infall of Ursa Major into Virgo is small and acts to increase Ursa Major's velocity (see Fig. 10 of

[Pierce & Tully, 1988]). However, if Virgo suffers from foreground contamination and is actually at a larger distance than Ursa Major, then this correction will be larger and of opposite sign, so the derived value of H_o would be significantly smaller. This suggests that the Ursa Major H_o estimate is not independent of the Virgo distance and re-emphasises the importance of Virgo for distance-scale studies.

4. Conclusions

From image sharpening observations at the William Herschel Telescope we have obtained 0.6-arcsec resolution images of the galaxies NGC4523 and IC3583. In the case of NGC4523 we have identified candidates for the brightest blue and yellow supergiants with R=21 mag in this galaxy. If the yellow supergiants are standard candles then this leads to an estimated distance modulus of $(m-M)_{AR}=30.6\pm0.3$ mag (11–15 Mpc) for NGC4523. If this corresponds to the Virgo distance then $H_o \approx 80$ kms^{-1}Mpc^{-1} and for an $\Omega_o=1$, inflationary Universe, the age of the Universe is uncomfortably short compared to the age of the oldest stars. However, there remains the possibility that NGC4523 and other line-of-sight spirals lie in the Virgo foreground, in which case substantially lower values of H_o may still be allowed. We are now making deeper, high resolution images of NGC4523 in an attempt to identify Cepheids and hence confirm this galaxy's distance. We are also attempting to find more brightest star distances to other Virgo galaxies.

References

[Cook et al., 1986] Cook K.H., Aaronson M., Illingworth G., *Astrophys. J., Vol. 301 (1986), p. L45.*

[Doel et al., 1990] Doel A.P., Dunlop C.N., Major J.V., Myers R.M., Purvis A., Thompson M.G., *Proc. SPIE, Vol. 19 (1990), p. 1236.*

[Freid, 1966] Fried D.L., *J. Opt. Soc. Amer., Vol. 56 (1966), p. 1372.*

[deLapparent et al., 1986] deLapparent V., Geller M.J., Huchra J.P., *Astrophys. J., Vol. 302 (1986), p. L1.*

[Harris, 1991] Harris W.E., *Ann. Rev. Astron. Astrophys., Vol. 29 (1991), p. 543.*

[Huchra, 1985] Huchra J.P., *in "The Virgo Cluster", Eds Richter O. & Binggeli B., Garching: ESO, 1985, p. 181.*

[Humphreys, 1988] Humphreys R.M., *in "The Extragalactic Distance Scale" Eds van den Bergh S. & Pritchet C.J., San Francisco: Astron. Soc. Pacif., 1988, p. 103.*

[Humphreys & Davidson, 1979] Humphreys R.M., Davidson K., *Astrophys. J., Vol. 232 (1979), p. 409.*

[Kraan-Korteweg et al., 1988] Kraan-Korteweg R.C., Cameron L.M., Tammann G.A., *Astrophys. J., Vol. 331 (1988), p. 620.*

[Pierce & Tully, 1988] Pierce M.J, Tully R.B. *Astrophys. J., Vol. 330 (1988), p. 579.*

[Pierce, 1991] Pierce M.J., *in "Observational Tests of Cosmological Inflation" Eds Shanks T. et al., Dordrecht: Kluwer Academic, 1991, p. 173.*

[Sandage & Bedke, 1985] Sandage A.R., Bedke J., *Astron. J., Vol. 90 (1985), p. 2006.*

[Sandage & Carlson, 1988] Sandage A.R., Carlson G., *Astron. J., Vol. 96 (1988), p. 1599.*

[Sandage & Tammann, 1990] Sandage A.R., Tammann G.A., *Astrophys. J., Vol. 365 (1990), p. 1.*

[Shanks et al., 1988] Shanks T., Tanvir N.R., Redfern R.M., *Publ. Astron. Soc. Pacif., Vol. 100 (1988), p. 1226.*

[Shanks et al., 1991] Shanks T., Tanvir N.R., Doel A.P., Dunlop C.N., Myers R.M., Major J.V., Redfern R.M., Devaney M.N., O'Kane P., *in "Observational Tests of Cosmological Inflation" Eds Shanks T. et al., Dordrecht: Kluwer Academic, 1991, p. 205.*

[Tanvir et al., 1991a] Tanvir N.R., Shanks T., Major J.V., Doel A.P., Myers R.M., Dunlop C., Redfern R.M., Devaney M.N., O'Kane P., *in "2nd Rencontres de Blois: Physical Cosmology", Ed. Tran Thanh Van J. et al., Editions Frontieres: France, 1991.*

[Tanvir et al., 1991b] Tanvir N.R., Shanks T., Major J.V., Doel A.P., Myers R.M., Dunlop C., Redfern R.M., O'Kane P., Devaney M.N., *Mon. Not. R. astr. Soc., Vol. 253 (1991), p. 21P.*

[Tully & Shaya, 1984] Tully R.B., Shaya E.J., *Astrophys. J., Vol. 281 (1984), p. 31.*

[vandenBerg, 1988] vandenBerg D.A., *in "The Extragalactic Distance Scale" Eds van den Bergh S. & Pritchet C.J., San Francisco: Astron. Soc. Pacif., 1988, p. 187.*

17

Adaptive Optics at ESO

Fritz Merkle *

Abstract

An adaptive optics system is one of the main features of the ESO-VLT – an array of four 8-m telescopes. These telescopes can be operated individually, in an incoherent, and in a coherent interferometric beam combination mode. Each telescope will be equipped with adaptive optics systems for real-time correction of atmospheric turbulence effects. First results with a prototype system developed for the VLT have now demonstrated the significant gain of this technology. This paper describes briefly the principles of adaptive optics, its planned application in the VLT project, and the status of its implementation programme.

1. Adaptive optics and the VLT

The VLT programme consists of an array of four 8-m telescopes arranged in a quatrilateral configuration. These four telescopes can be operated individually, in a so-called incoherent combined mode (mainly for high-resolution spectroscopy), and in a coherently combined mode as a long-baseline interferometer to achieve highest spatial resolution. In order to increase the performance and use of the VLT in the interferometric mode, two to four mobile telescopes of the two-metre class will be added to the array. The gain of large telescopes for astronomical spatial interferometry depends substantially on the availability of adaptive optics, allowing diffraction-limited imaging of each individual telescope in the near infrared and a partial correction of atmospheric distortions at visible wavelengths [Merkle, 1988]. In addition, spectroscopy would profit significantly from adaptive optics, because the light flux can be concentrated on a much narrower entrance slit of the spectrograph, which increases the spectral resolution and allows design of a more compact instrument. Finally, single-aperture diffraction-limited imaging would give each telescope an unprecedented performance.

Therefore, it is planned to equip all telescopes of the VLT with adaptive optics systems [Hardy, 1978; Pearson *et al.*, 1979; Merkle, 1991] with the goal to achieve

---

*European Southern Observatory, Karl-Schwarzschild-Strasse 2, D-08046 Garching, Federal Republic of Germany.

diffraction-limited imaging for wavelengths greater than 2.2 μm and partial correction at shorter wavelengths. Adaptive optics techniques are closely related to *active optics*, a main feature of the VLT unit telescopes. Both techniques apply a real-time closed-loop correction to the wavefronts. In the case of active optics, it is the compensation of wavefront aberrations due to primary mirror and telescope effects originating from polishing, gravity, temperature, wind buffeting, and others. The corrections are applied directly to the relatively flexible primary mirror and the steerable secondary mirror. In the case of *adaptive optics*, wavefront aberrations due to optical propagation effects in the atmosphere above the telescope optics from inside the telescope enclosure all the way up to approximately 10 km are compensated in real-time with an additional optical element, a small deformable mirror. A major difference between these two applications of the same physical principle is in the control bandpass. For active optics the typical frequency ranges are DC to 1 Hz and for adaptive optics 1 Hz to 200 Hz or even higher.

2. Principles of adaptive optics

It is a well known and widely accepted fact that the imaging quality of ground-based telescopes is degraded by the transmission of the light from the astronomical object through the turbulent atmosphere [Roddier, 1981]. The reason for this degradation is a random spatial and temporal wavefront perturbation induced in the different layers of the atmosphere. In addition, the residual aberrations of the optical elements of the telescope contribute to the degradation. All these wavefront perturbations together result in a complex phase aberration

$$\Phi(r, t) - iA(r, t).$$

The real part $\Phi(r, t)$ represents the phase-shift of the wavefront, usually called "seeing", while the imaginary part $A(r, t)$ is a measure of the intensity fluctuations across the aperture plane, called "scintillation".

These phase-shifts will be corrected by an adaptive optics system. It uses a phase-shifting optical element, a deformable mirror, which can be controlled in space and time in order to compensate the atmospheric phase shifts. A system for the correction of these aberrations applies then a compensation equal to

$$-\Phi(r, t) + i\mu A(r, t)$$

where μ is a dimensionless intensity scaling factor. For most of the imaging problems, especially with very large telescopes, the phase-correction part is fully sufficient.

The first suggestions for the construction of an adaptive optical correction device to solve the astronomical seeing problems originated in 1953 [Babcock, 1953]. The technology of adaptive optics has been developed during the past 15 to 20 years mainly for laser propagation applications. In October 1989, it was first demonstrated [Rousset *et al.*, 1989; Merkle *et al.*, 1990a; Merkle *et al.*, 1990b], that adaptive optics also works reliably for the correction of astronomical infrared images. These

214

Table 1. Parameters for adaptive optics for the VLT

λ (μm)	0.5	2.2	5.0	10
r_0 (cm)	10	60	160	360
N	6400	180	12	4
τ (ms)	6	35	95	220
Θ (arcsec)	1.8	10	30	70

experiments have been carried out with the VLT adaptive optics prototype system (also called the "COME-ON" system) on the 1.52-m telescope of the Observatoire de Haute Provence in a collaboration of the European Southern Observatory with the Observatoire de Paris, ONERA, and Laserdot (formerly Laboratoire de Marcoussis). Meanwhile these experiments have been successfully repeated on the ESO 3.6-m telescope and first astronomical observing programmes have taken place.

3. General parameters for an adaptive optics system

The number of subapertures (N), the correction rate ($1/\tau$), and the wavefront correction range (Δz) are the basic parameters for an adaptive optics system; typical values of these parameters at different wavelengths are given in Table 1 for the VLT case. (A seeing with a Fried parameter of 10 cm is assumed.) Δz is in the range ± 12.5 μm.

For an 8-m telescope this would lead to more than 6000 controlled subapertures and correction rates higher than 170 Hz. A realistic aim for adaptive optics systems which will be available in the early phase of the VLT operation, is full wavefront compensation at infrared wavelengths (≥ 2.2 μm) with a system of approximately 250 subapertures and a correction rate of higher than 50 Hz.

4. Elements of an adaptive optics system

The main elements of an adaptive optics system are the wavefront correction device, the wavefront sensor, and the control computer (see Fig. 1).

For the correction of the atmospheric perturbations in an astronomical telescope the continuous thin-plate mirrors with discrete position actuators or bending-moment actuators (as in bimorph mirrors [Roddier *et al.*, 1991]) seem to be the most favourable.

In astronomy with the severe intensity problems due to the nature of the objects, a Shack–Hartmann wavefront sensor seems to be the most adequate. The Shack–Hartmann sensor is based on the well-known Hartmann test for checking the figure

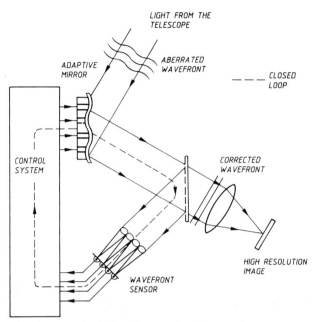

Figure 1. Principle of an adaptive optics system.

of large optical elements. At ESO the Shack–Hartmann sensor was for the first time implemented in 1979 [Noethe *et al.*, 1984]. Shack–Hartmann type sensors are unique when compared to other systems because they collect and sample virtually 100% of the light entering the optical system. Additionally, the wavefront tilt over the subapertures can be measured even when the phase of the light from one side of the subaperture to the other side exceeds 2π. In addition, Shack–Hartmann type sensors work for broadband light.

The control algorithms will be based on modal correction. Modal-correction algorithms have the advantage of allowing partial correction of a limited number of low spatial-frequency modes, e.g. in cases where the high spatial-frequency components of the wavefront cannot be measured with sufficient signal-to-noise ratio due to the faintness of the source or to turbulence which is too strong. These algorithms require very high computational power in order to meet the temporal and spatial requirements. With special dedicated hardware or hybrid systems this problem has been approached successfully. Recently, neural networks have also been successfully employed for this purpose [Angel *et al.*, 1991].

5. The VLT adaptive optics prototype and first results

Based on these considerations a prototype system has been developed and tested, as mentioned above. It is based on a 19-actuator deformable mirror with discrete piezoelectric actuators. The stroke of each actuator is ±7.5 μm. Global wavefront tilt

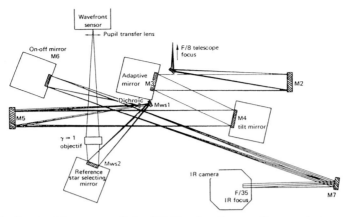

Figure 2. Optical lay-out of the VLT adaptive optics prototype system.

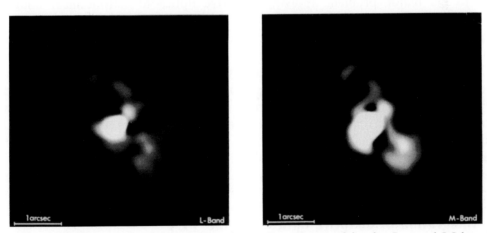

Figure 3. Diffraction limited image of Eta Car observed in the L-, and M-band.

correction is separated from the higher orders of aberration by using an additional tip/tilt mirror. The wavefront sensor is of the Shack–Hartmann type with 5×5 subapertures. An electron-bombarded CCD makes it photon-noise limited. The computing power for the control system comes from a dedicated hardware device with a 68020 processor based host computer. The system is equipped with a 32×32 IR-array camera. Fig. 2 shows the layout of the system [Merkle *et al.*, 1990b]. Currently this system is being upgraded by installing a 52-actuator mirror, a 7×7 subaperture wavefront sensor, and a faster control computer.

To illustrate the performance of the system one of the first scientific results is shown in Fig. 3 [Rigaut *et al.*, 1992].

Table 2. Limiting magnitudes and sky coverage for adaptive optics

λ (μm)	0.5	2.2	5.0	10
r_0 (cm)	10	60	160	360
m_{lim} (mag)	7	13	15.5	17
C_P (%)	$\simeq 0$	0.1	30	100
C_E (%)	$\simeq 0$	0.3	100	100

6. Limitations of an adaptive optics system

Any correction requires a measurement of the effect which will be corrected. This leads to the major problem applying adaptive optics to astronomical imaging. The observed sources are in most cases so faint that their light is insufficient for a wavefront determination. A brighter nearby reference source within the same isoplanatic patch is rarely available.

The sky coverage with bright enough reference objects is not sufficient for the operation of an adaptive optical system at visible wavelengths for any source. Under the assumption that a minimum of approximately 100 photons per subaperture are required for a wavefront measurement and with integration times of less than 10 msec, the limiting magnitude m_{lim} would be unsatisfactory by far. However, for infrared wavelengths the situation is more favourable because of the increase of r_0, τ, and isoplanatic angle. Table 2 gives the resulting limiting magnitude and sky coverage for some selected wavelengths.

7. Integration of adaptive optics into the VLT

The VLT adaptive optics systems will be integrated into the coudé trains of the 8-m unit telescopes (see Fig. 4). The deformable adaptive mirror is the fifth optical element (M5) in the coudé path. At this location is an intermediate pupil with a diameter of 120 mm. The wavefront sensor will be located at the individual coudé focus behind the dichroic mirror which deflects the light to a high resolution infrared camera or to the interferometric beam combination. In case adaptive optics are not used there are two options, either switching to a flat passive mirror at position M5, or keeping the deformable mirror, which could provide in this case a DC wavefront correction for all constant residual aberrations of the coudé path.

The system will be composed of subapertures of size 40×40 cm or slightly less depending on the mirror technology available in the future. This leads to a deformable mirror with approximately 250 actuators with a stroke of approximately ± 4 μm. This is sufficient because this mirror will be mounted in a tip/tilt support

218

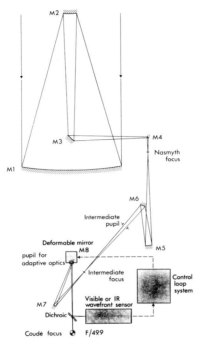

Figure 4. Optical lay-out of the coudé path of the VLT and the location of the deformable mirror and wavefront sensor.

in order to separate the tilt correction from the higher modes. The tilt mode has the highest contribution to the total wavefront error.

The wavefront sensor will be based again on the Shack–Hartmann principle. The wavefront will be sampled by the microlenses in a raster of approximately 20×20 subapertures. It is planned to mount the wavefront sensor on a scanning table in order to locate it at any place within the coudé field (± 1 arcmin in any direction). The science object will always stay on-axis. It would be used as the reference source when it is bright enough for wavefront measurement. The wavefront will be measured in the visible range and as an option, with an additional infrared wavefront sensor in the H-, J-, and K-band. Both devices would be mounted permanently in order to switch the sensing wavelength in a very short time.

A user interface will be provided in order to operate the adaptive optics system as a common-user instrument. Automated as well as manual reference source selection will be provided. A real-time output of atmospheric turbulence data such as seeing, coherence time, and other parameters may be provided. On-line status data such as correction quality and expected Strehl ratio with and without correction are features foreseen when routine operation starts.

8. Operation modes of the VLT adaptive optics systems

The adaptive optics systems for the VLT will provide the following operation levels:

The lowest level will be a pure image-motion correction. In this case the deformable mirror will compensate only the DC aberration of the coudé train by applying a constant off-set to the mirror actuators.

The highest level of correction will be when the system corrects all aberrations within the range of the spatial and temporal resolution of the system. This also sets the highest demands on the brightness of the reference source.

The adaptive optics system will provide as an additional feature any correction between pure tilt correction and full correction up to the highest mode. By changing the sampling of the wavefront (number of subapertures) it is possible to increase the sensitivity of the system at the cost of level of correction.

In addition it will be possible to operate only the wavefront sensor and record wavefront sequences synchronously with the integration sequence of the IR camera for a later off-line correction of the speckle images or for image-selection procedures.

9. Expected performance of the systems

Based on the experience gained with the VLT Adaptive Optics Prototype System the following performance of the systems with which the VLT will be equipped is expected:

Full correction (Strehl ratio $\geq$ 80% of the theoretical value) for wavelength greater 2.2 μm.

Sky coverage for full correction in the L-band will be close to 100% in the Galactic plane and in the range of 30% close to the Poles (assuming 10-cm seeing at 0.5 μm). These values will depend on the sensitivity and wavelength range of the wavefront sensor. For longer wavelengths there will be no restriction in coverage. At shorter wavelengths the coverage is limited. In these cases the correction can be restricted to lower modes, which will increase the application range but at the cost of even more limited partial correction.

The correction is best in the close neighborhood of the reference source and degrades continuously with increasing distance inside the isoplanatic patch. Therefore it is preferable to use the science object as reference if possible.

Image motion stabilization will be better than 0.05 arcsec.

220

10. The VLT adaptive optics programme

The 19-actuator adaptive optics prototype system was a first step into the field of adaptive optics for the ESO-VLT. It serves as a testbench for the various elements of an adaptive optics system and the control algorithms. In a second development phase, started in the third quarter of 1990 the prototype system will be upgraded as mentioned above. It will then serve as the intermediate-scale system towards the large units required for the VLT already featuring some of the technology to be applied later.

The beginning of design and construction of the four systems for the 8-m telescopes as well as the two-meter class telescopes is planned for the second half of 1993. The first system should then be ready for installation at the telescope by end of 1996.

11. Future features for adaptive optics of the VLT

New techniques to overcome the reference-source problem even in the visible range have been proposed [Foy & Labeyrie, 1985]. With a LIDAR-like technique an artificial reference source can be generated [Thompson & Gardner, 1987] by using resonance scattering of yellow laser light in the mesospheric sodium layer of 80 to 100 km height, or by Rayleigh scattering of laser light of shorter wavelength between 5 and 15 km above ground. These artificial reference sources or laser guide stars will have to be generated within the isoplanatic angle of the astronomical source at a repetition rate synchronous with the adaptive correction rate. It will be necessary to gate the astronomical detector during the wavefront sensing time, which means a small loss compared to the high gain of the adaptive correction. This technique has already been successfully tested for military applications which clearly demonstrated the feasibility [Fugate et al., 1991; Primmerman, 1991]. The test of this technique for astronomy is expected to be included in further investigations of adaptive optics for the VLT. With the availability of laser guide stars it can be expected that full sky coverage down to the visible-wavelength range can be obtained.

12. Conclusions

The ESO activities in the area of adaptive optics have demonstrated clearly the potential of this technique for astronomical research. All four 8-m telescopes of the VLT will be so equipped. It is expected that within the next few years adaptive optics will become standard equipment for the larger telescopes at major observatories.

References

[Angel et al., 1991] Angel J.R.P., Wizinowich P., Lloyd-Hart M., Sandler D., *Nature, Vol. 348 (1991), p. 221.*

[Babcock, 1953] Babcock H.W., *Publ. Astron. Soc. Pacif., Vol. 65 (1953), p. 229.*

[Foy & Labeyrie, 1985] Foy R., Labeyrie A., *Astron. Astrophys., Vol. 152 (1985), p. L29.*

[Fugate *et al.*, 1991] Fugate R.Q. *et al.*, *Nature, Vol. 353 (1991), p. 144.*

[Hardy, 1978] Hardy J.W., *Proc. IEEE, Vol. 66 (1978), p. 651.*

[Merkle, 1988] Merkle F., *J. Opt. Soc. Amer. A, Vol. 5 (1988), p. 904.*

[Merkle, 1991] Merkle F., *Adaptive Optics. in International Trends in Optics, Ed. Goodman J., Academic Press: Boston, 1991.*

[Merkle *et al.*, 1990a] Merkle F., Gehring G., Rigaut F., Kern P., Gigan P., Rousset G., Boyer C., *The Messenger, Vol. 60 (1990), p. 9.*

[Merkle *et al.*, 1990b] Merkle F., Rousset G., Kern P., Gaffard J.P., *Proc. SPIE, Vol. 1236 (1990), p. 193.*

[Noethe *et al.*, 1984] Noethe L., Franza F., Giordano P., Wilson R.N., *Proc. IAU Colloq. 79 (1984), p. 67.*

[Pearson *et al.*, 1979] Pearson J.E., Freeman R.H., Reynolds H.C., in *Applied Optics and Optical Engineering, VII, Academic Press, 1979.*

[Primmerman, 1991] Primmerman C.A., *Nature, Vol. 353 (1991), p. 141.*

[Rigaut *et al.*, 1992] Rigaut F., Rousset G., Kern P., Fontanella J.C., Gaffard J.P., Merkle F., Léna P., *Astron. Astrophys., (1992), in press.*

[Roddier, 1981] Roddier F., *The Effect of Atmospheric Turbulence in Optical Astronomy. in Progress in Optics, Ed. Wolf E., North Holland, 1981.*

[Roddier *et al.*, 1991] Roddier F., Northcott M., Graves J.E., *Publ. Astron. Soc. Pacif., Vol. 103 (1991), p. 131.*

[Rousset *et al.*, 1989] Rousset G., Fontanella J.C., Kern P., Gigan P., Rigaut F., Léna P., Boyer C., Jagourel P., Gaffard J.P., Merkle F., *Astron. Astrophys., Vol. 230 (1989), p. L29.*

[Thompson & Gardner, 1987] Thompson L.A., Gardner C.S., *Nature, Vol. 328 (1987), p. 229.*

18
The challenge of excellent sites

René Racine *

Abstract

Excellent ground-based sites deliver sub-half-arcsec natural seeing (median) and frequently reach the sub-quarter-arcsec level. Experience gained at the CFH Telescope on Mauna Kea explains why such performances are seldom realized. New levels of excellence in the design, figuring, implementation and management of astronomical optics are needed and can usher in a revolution in ground-based astronomy. Simple scaling laws are derived and used to forecast the imaging performance of telescopes as a function of their primary mirror aperture and f-ratio. It is speculated that future, not-so-very-large telescopes, highly optimized for image quality, will occupy a privileged niche in twenty-first-century astronomy.

1. Introduction

Quantitative *in situ* studies of atmospheric turbulence [Hufnagel, 1974; Barletti *et al.*, 1976; Bely, 1987; Vernin & Azouit, 1987; Roddier *et al.*, 1990] and experience with large telescopes at excellent sites, notably in Chile, Hawaii and the Canary Islands, has slowly made the astronomical community aware that the ~ 1 arcsec norm of "seeing" excellence, which prevailed twenty years ago, could be much surpassed with optimized facilities at such sites. I will demonstrate that, to take full advantage of the natural image quality offered by excellent sites, the global facility error budget for imaging must be < 0.1 arcsec. This poses tall technological challenges, as experience at the Canada-France-Hawaii Telescope (CFHT) and elsewhere shows.

The scientific benefits of meeting these challenges stems from two fundamental considerations : With a telescope of aperture D delivering a PSF of spread angle ω, the photometric signal-to-noise ratio S/N in a sky-limited unresolved image scales as S/N $\propto D/\omega$. Since the number of resolvable elements in a 2-D frame scales as $1/\omega^2$, the "structural S/N" is proportional to $1/\omega$. Thus the global figure of merit of a telescope for 2-D studies of faint sources, a very common exercise in astronomy, can be measured by the quantity $Q=D/\omega^2$. The interest of improved image quality (IQ) becomes quantitatively obvious!

---
*Département de Physique, Université de Montréal, and Observatoire du Mont Mégantic, Montréal, Québec, Canada H3C 3J7.

Table 1. Free atmosphere spread angles

Altitude (m)	$\langle FWHM \rangle$ (arcsec)	$\langle FWHM \rangle_s$ (arcsec)	$FWHM_s$, 10% (arcsec)
200	1.30	1.30	0.82
1000	0.70	0.65	0.48
2000	0.52	0.44	0.25
4000	0.42	0.30	0.18

2. Things should be better than they are

Table 1 gives the spread angles (FWHM, arcsec) expected from turbulence in the free atmosphere for sites at various altitudes above sea level, and the corresponding median and 10 percentile angles for stabilized imaging with a 3.6-m telescope. Stabilized imaging removes image wander due to the tilt components of the atmospheric aberration; this is easily achieved with a "fast guider" capable of performing at ~ 10 Hz.

One sees that at high-altitude isolated sites, where the planetary boundary layer is weak, telescopes located above the ground-induced microthermal turbulence should normally achieve sub-half-arcsec imaging, and simple "fast guiding" can bring sub-quarter-arcsec images frequently within reach of large instruments.

Unfortunately, as all astronomers know, reality is seldom that bright. This is so because, to take full advantage of nature's generosity, the global error budget of the facility must be < 0.1 arcsec! This would increase ω by $< 20\%$ at the 10 percentile and by $< 10\%$ at the median. Therein lies the challenge of excellent sites to optical designers, opticians, telescope engineers, observatory managers and astronomers.

Fig. 1 shows the histogram of the image quality (FWHM arcsec) actually achieved with HRCam, a fast guider/imager at the CFHT [McClure *et al.*, 1989]. The median of 0.62 arcsec is poorer than that expected from Table 1 by a factor of ~ 2. This means that for morphological studies the number of resolved elements is reduced by a factor of 4, and that the 3.6-m CFHT detects sky-limited unresolved sources with an efficiency no better than that of a "perfect" 1.8-m telescope at the site.

3. Things used to be worse

Fig. 2 shows the historical evolution of IQ at the CFHT. During its first year of operation (1980), with $\langle IQ \rangle \sim 2.0$ arcsec, the 3.6-m CFHT was, for many purposes, equivalent to a "perfect" 0.5-m telescope! Most of the IQ improvement has come from the elimination of "dome seeing" due to thermal inhomogeneities around the telescope. The periodic peaks of poorer IQ during the first quarters of each year, when the outside air temperature is the coldest, show that extraneous heat

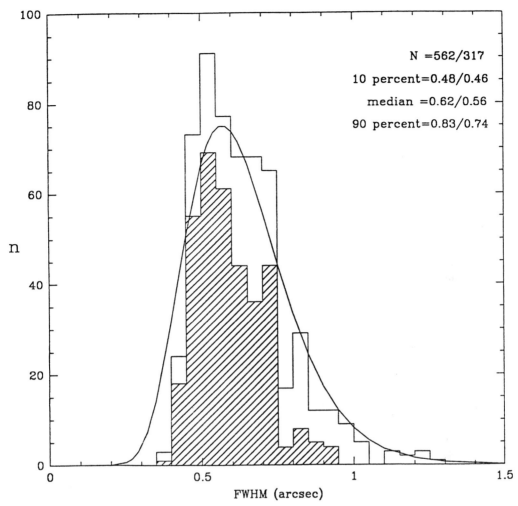

Figure 1. Histogram of stabilized image quality achieved with HRCam at the CFHT. The global median (no selection) is 0.62 arcsec FWHM.

sources were largely responsible for the IQ deterioration until $\sim$ 1988. In recent years, efforts at the CFHT have concentrated on the identification and correction of residual optical aberrations, on quantitative studies of specific local seeing problems such as "mirror seeing", and on the development of instrumentation capable of partial correction for atmospheric turbulence, HRCam being the simplest and first to be commissioned.

A different display of the CFHT IQ data, suggested to me by G. Jacoby, and which vividly illustrates the gains made over the years in fighting "local seeing", is shown in Fig. 3. The transformed ordinate is now $1/(\omega^2 - 0.4^2)$, i.e. the telescope

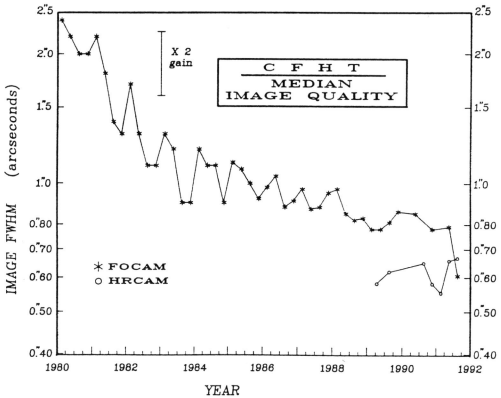

Figure 2. Historical evolution of image quality at the CFHT. The recurrent peaks during the (colder) first quarters show that until ~ 1988 heat leaks to the observing area were responsible for image deterioration.

efficiency ($\propto \omega^2$)as affected by local seeing but corrected for the known spread of ~ 0.4 arcsec due to the optics (see §4). The linear growth of the CFHT efficiency, which started at $\sim$ zero at its commissioning in September 1979 and is still rising more than a decade later, is a testimony to the continuing success of CFHT staff in perfecting the facility. Note, however, that since the median natural seeing (unstabilized) on Mauna Kea is $\langle \omega_n \rangle \simeq 0.45$ arcsec (FWHM) [Racine *et al.*, 1991] the efficiency limit is $1/\langle \omega_n \rangle^2 = 5$. With the historical linear growth rate of Fig. 3, another decade of efforts would be needed to attain that limit.

4. Things can be made better

We now know what still remains to be improved at the CFHT to sharpen IQ further. Fig. 4 demonstrates that "mirror seeing" can be important. When the primary mirror is warmer than ambient air, image spread increases at the rate of ~ 0.4 arcsec (FWHM)/°C. To avoid this single item from gobbling up the full

226

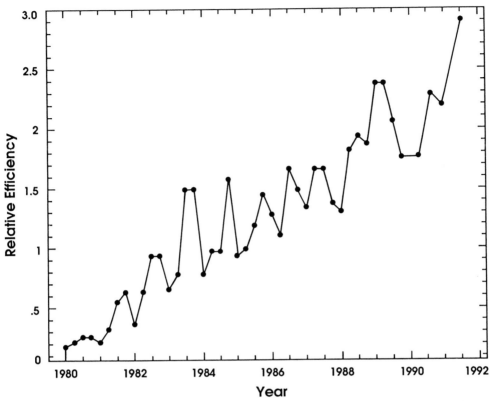

Figure 3. Same as Fig. 2, but transformed to show the evolution of the telescope efficiency, $1/(\omega^2 - 0.4^2)$. See text.

allowable error budget of 0.1 arcsec demands $\Delta T(\text{mirror}) < +0.1°C$, a respectable challenge indeed. It is in fact doubtful that this challenge can be consistently met. One must then invent ways to prevent air convection from developing above the mirror surface, by flushing the surface with a laminar air flow for instance, a suggestion which goes back more than 30 years [Steavenson, 1960; Hysom, 1960].

Exacting telescope focusing is a frustrating art in this era of enormous pressure on large telescope time. For image spread due to a focus error Δf to be <0.05 arcsec requires $\Delta f/f < 2.4 \times 10^{-7}$ (f/D), or $\Delta f < 12$ μm for a 3.6-m f/3.8 primary such as the CFHTs. With a steel telescope tube, $\Delta f/f \sim 1 \times 10^{-5} \cdot \Delta T$ (°C). Focus needs to be adjusted when $|\Delta T| > 0.025 \cdot$ (f/D)°C. On Mauna Kea, where the night air temperature is remarkably stable ($\Delta T/t > \sim 0.3$ °C/hr), focusing with the relatively slow primary CFHT should still be adjusted every ~ 20 min on the average. Since it is not currently practical to do this, focus spread contributes 0.1–0.2 arcsec of image spread on the average at such telescopes. With the much faster optics planned for the very large telescopes (VLTs), and at sites where air temperature varies more

227

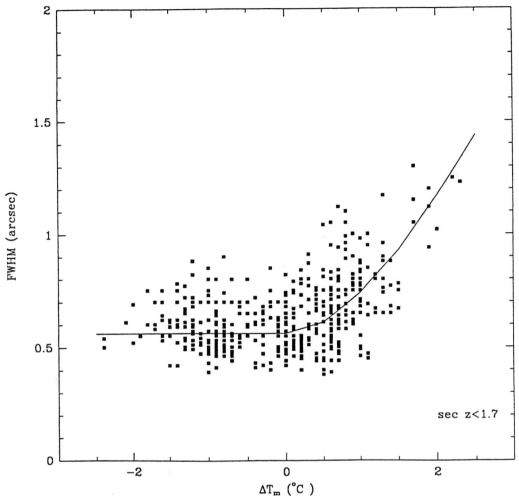

Figure 4. "Mirror seeing" at the CFHT. The line corresponds to a constant IQ to which a degradation of 0.4 arcsec/ °C is added for $\Delta T > 0$°C.

rapidly, focusing "by the minute" will be a necessity. An f/1.5 8-m primary requires $\Delta f < 4$ μm to keep focus spread below 0.05 arcsec. At that level, the structure of an 8-m telescope behaves like jelly and active focusing (and collimation, and mirror figure control) becomes an absolute requirement.

Fig. 5 shows how the HRCam "seeing" varies with zenith distance. A detailed analysis of these data is given by Racine *et al.* [1991]. A least-squares solution yields a slope corresponding to a median zenith seeing of 0.32 arcsec FWHM, much as Table 1 would predict. The actual intercept at the zenith can be reconciled with this

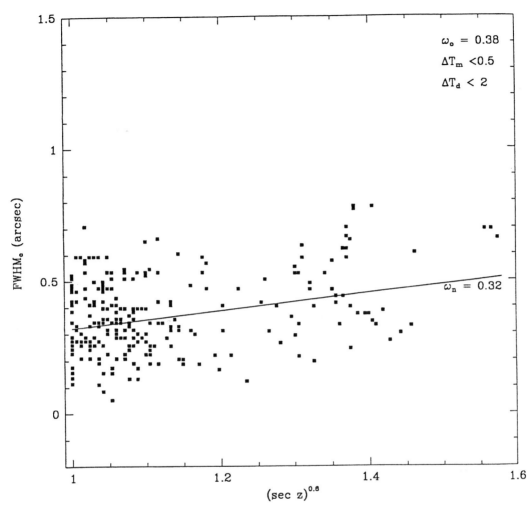

Figure 5. HRCam IQ as a function of zenith distance. The slope yields a stabilized natural seeing of 0.32 arcsec in the zenith. The intercept requires a further contribution of $\sim$ 0.38 arcsec from the optics.

slope if a typical contribution of 0.38 arcsec to image spread is attributed to optical aberrations, of which defocus is but one.

The wavefronts from the CFHT primary mirror for various attitudes of the telescope are shown in Fig. 6. Defocus and coma were removed from these Shack–Hartmann generated maps. The FWHM of the PSF resulting from these residual aberrations is $\sim$ 0.2 arcsec, well within the original specifications the Scientific Committee challenged the opticians to meet in 1975. But what was fine by the 1975 standards of expected natural seeing is no longer sufficient. Higher-order aberrations, notably spherical astigmatism, are due to figuring and to slight imperfections of the

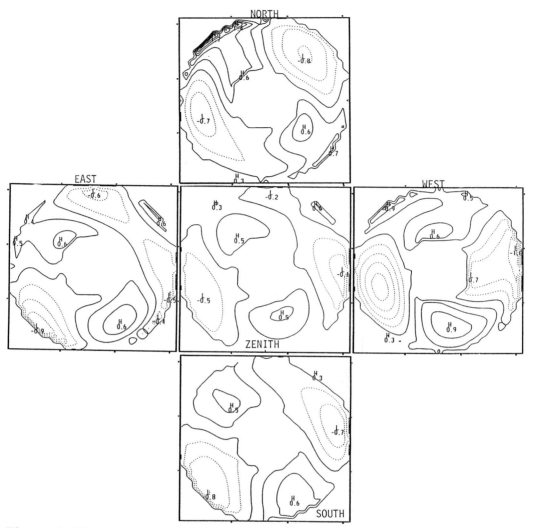

Figure 6. Wavefront aberrations (defocus and coma removed) for different tele-scope attitudes. Variations are apparent. On the average, the resulting PSF has FWHM ∼ 0.2 arcsec.

mirror support system, which underwent a major readjustment in 1988. Some of these aberrations would be correctable with active optics in today's new-technology telescopes. A phase-corrector plate will be installed at a pupil in HRCam to remove the fixed components of mirror imperfections.

Mechanical flexures introduce decollimation coma at the CFHT, as illustrated in Fig. 7 by the variable decentring of the central shadow in these defocused images. At a zenith distance of ∼ 60 deg, the full length of the coma figure is ∼ 0.7 arcsec. This corresponds to ∼ 3.5 mm of decentring and to an effective image spread of

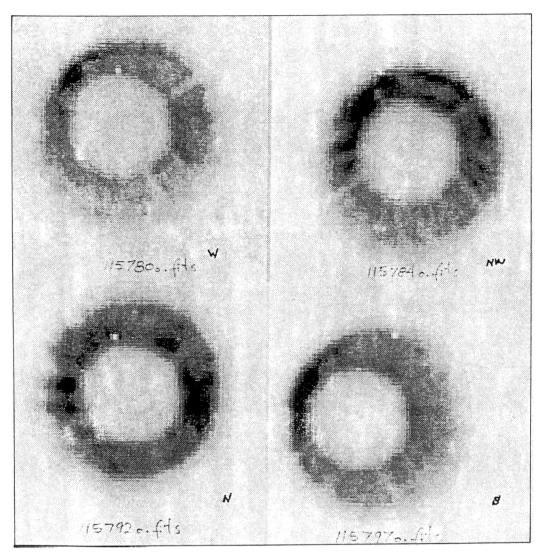

Figure 7. Defocussed HRCam images for different telescope attitudes. The decentring of the central shadow reveals coma introduced by variable telescope flexure and decollimation.

~ 0.25 arcsec FWHM. Wilson *et al.* [1987] found similar amounts of coma at the ESO 3.6-m. Since decentring appears to be quite repeatable at the CFHT, the proposed cure is to motorize the three defining pads of the primary mirror and to correct decollimation in an open-loop fashion.

For a telescope of aperture D affected by a decentring Δp, decollimation coma is proportional to $\Delta p/[D \cdot (f/D)^3]$. For the angular value of the coma figure to be < 0.1 arcsec with an f/1.8, 8-m primary, the collimation of the prime focus instruments or secondary mirrors must be controlled to better than 120 μm. That tight tolerance would be further reduced to 35 μm at f/1.2. Although wavefront sensors should certainly be capable of detecting the coma to be corrected, by the displacement or tilt of the secondary mirror for instance, considerable attention will be required in the design, construction and maintenance of the mechanisms needed to locate tons of opto-mechanical elements to this level of accuracy.

The instruments themselves add to the image spread introduced by the telescope. For instance, we know that the design and production of the rather complex HRCam optics introduce $\sim$ 0.15 arcsec of image spread. To quote Charles Wynne: "The best optics have the least optics."

It should now be clear that the $\sim$ 0.4 arcsec of optics spread required by the intercept at the zenith in Fig. 4 can indeed be accounted for by documented components of the actual CFHT error budget:

defocus	$\sim$ 0.15 arcsec
mirror figure	$\sim$ 0.20 arcsec
coma	$\sim$ 0.25 arcsec
HRCam optics	$\sim$ 0.15 arcsec

The corollary is that the actual median IQ delivered by the CFHT site, after tip-tilt corrections by HRCam, is indeed $\sim$0.30 arcsec as computed from the Barletti *et al.* [1976] free atmosphere (Table 1).

The actual frequency distribution of the tilt- and optics-corrected zenith IQ at the CFHT is shown in Fig. 8. The median and 10 percentile are at 0.32 arcsec and 0.19 arcsec respectively. These should be taken as typical performances to be expected from carefully designed, built and managed astronomical facilities at excellent sites. A few decades ago, when current 4-m class telescopes were on the drawing boards, setting a global IQ error budget of < 0.1 arcsec would have seemed to be a totally preposterous goal. Now, this should be regarded as necessary to meet the challenge of excellent sites and to exploit fully their potential for astronomy, however demanding that specification may appear to be from an engineering point of view.

5. A new breed of telescope

It is clear that new technologies, fastidious instrumental care and quite strict operational policies are needed to reap the full benefits ground-based telescopes can deliver. Such telescopes would then bring a wealth of novel information and scientific results, and can usher in a new era in ground-based astronomy, not unlike the one expected from major space facilities.

However, one easily suspects that, for a given technology, the actual imaging performances of a telescope will depend on the parameters of its optical design. This is so because residual imperfections (optical figure, flexure, etc.) will always be

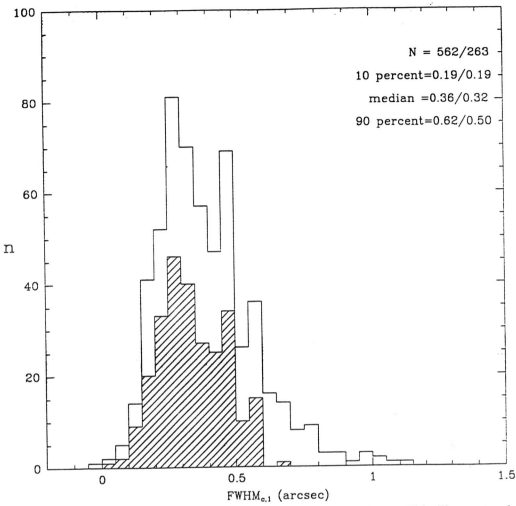

Figure 8. Optics-corrected IQ histogram from HRCam data. This illustrates the performance goals to be met by optimized telescopes at excellent sites.

present and because their effects on IQ are related to such parameters as sizes and f-ratios. Let us summarize the scaling laws involved.

Diffraction introduces a spread $\omega_d(\text{arcsec}) = 0.2\lambda(\mu m)/D(m)$

Optical figure errors should be proportional to how much the optical surface deviates from a sphere: $\omega_o \sim D/(f/D)^3$. From the actual performance of "old technology" mirrors, one can put ω_o (arcsec) $= 3D(m)/(f/D)^3$. This yields 0.2 arcsec for an f/3.8, 3.6-m primary, as is actually obtained at the CFHT, and 14 arcsec for an f/1.2 8-m primary.

233

Table 2. Expected IQ (arcsec) in 0.2 arcsec seeing; new technology: 10×improvement; $\lambda = 0.7\mu$m

| f/D | 3.0 | | 1.8 | | 1.2 | |
D (m)	ω	Q	ω	Q	ω	Q
2	0.21	43	0.24	34	0.4	11
4	0.21	89	0.31	42	0.8	7
6	0.22	124	0.40	38	1.1	5
8	0.23	150	0.50	32	1.5	4
10	0.25	164	0.61	28	1.8	3

Focus error (see above) typically produces ω_f (arcsec) $= 0.6/(f/D)$.

For decollimation coma, $\Delta p \sim W \cdot l/k$, where W is the deflecting force (weight $\propto$ D^3), l the linear length ($\propto$ f) and k is the rigidity of the tube structure ($\propto D^2$). For "old technology" telescopes, this predicts ω_c (arcsec) $= 0.8 \cdot D(m) / (f/D)^2$.

These are the dominant terms for optics spread, assuming perfect guiding. They should be combined ($\sim$ quadratically) with the image spread due to the atmosphere, ω_n, to estimate the global image spread. The coefficients of these scaling laws can be lowered by new technologies in optical figuring and telescope engineering, including adaptive optics. One could, optimistically perhaps, assume a tenfold improvement in all factors (except diffraction!) to forecast the imaging performance of new-technology telescopes. Table 2 gives the result of this exercise with 10th percentile natural images ($\omega_n = 0.2$ arcsec). The figure of merit $Q \equiv D/\omega^2$ is given for each case.

The results highlight the extreme sensitivity of the global IQ to the primary f-ratio. Since very large telescopes must have a fast primary (f/D $<$ 2), to be accommodated in reasonably sized enclosures, even a tenfold improvement in each and every one of the performance parameters will fail by a large margin to meet the challenge of excellent sites. For the same 10 $\times$ improvement, and with fast primaries, smaller telescopes have a higher IQ than larger ones. For an 8-m telescope, gains of 50 $\times$ will be needed at f/1.8 and of 200 $\times$ at f/1.2 to produce sub-quarter-arcsec images. It may well be that, in the decades ahead, the ultimate in angular resolution and detection against sky noise will only be achievable with telescopes whose sizes are such that they can benefit from a vast operational experience making them ripe for optimization, and which do not require stupendous engineering feats to ensure their ultimate performances. While the light-gathering power of 8 to 10-m VLTs will make them uniquely powerful for high S/N work on bright stars (where S/N $\propto$ D^1) or on featureless sources (S/N $\propto D^0$), in the realm of faint stellar sources (S/N $\propto D/\omega$) or of resolved objects (S/N $\propto D/\omega^2$), 2 to 4-m new-technology telescopes should remain unsurpassed for decades to come.

References

[Barletti *et al.*, 1976] Barletti R., Ceppatelli G., Paternò L., Righini A., Speroni N., *J. Opt. Soc. Amer.*, Vol. 66 (1976), p. 1380.

[Bely, 1987] Bely P.-Y., *Publ. Astron. Soc. Pacif.*, Vol. 99 (1987), p. 560.

[Hufnagel, 1974] Hufnagel R.E., in *Topical Meeting on Optical Propagation through Turbulence, University of Colorado: Boulder, 1974.*

[Hysom, 1960] Hysom E.J., *J. Brit. astr. Assoc.*, Vol. 71 (1960), p. 274.

[McClure *et al.*, 1989] McClure R.D., Grundmann W.A., Rambold W.N., Fletcher J.M, Richardson E.H., Stillburn J.R., Racine R., Christian C.A., Waddell P., *Publ. Astron. Soc. Pacif.*, Vol. 101 (1989), p. 1156.

[Racine *et al.*, 1991] Racine R., Salmon D., Cowley D., Sovka J., *Publ. Astron. Soc. Pacif.*, Vol. 103 (1991), p. 1020.

[Roddier *et al.*, 1990] Roddier F. *et al.*, *Proc. SPIE*, Vol. 1237 (1990), p. 336.

[Steavenson, 1960] Steavenson W.H., *J. Brit. astr. Assoc.*, Vol. 70 (1960), p. 206.

[Vernin & Azouit, 1987] Vernin J., Azouit M., *ESO VLT Report No. 55 (LASSCA),* 1987, p. 7.

[Wilson *et al.*, 1987] Wilson R.N., Franza F., Noethe L., *J. Mod. Opt.*, Vol. 34 (1987), p. 475.

19

Testing telescopes from out-of-focus images: Application to ground-based telescopes and to the Space Telescope

Claude and François Roddier [*]

Abstract

It is shown that the optical quality of astronomical telescopes can be quantitatively tested by simply analyzing a small set of properly defocused images. Two types of algorithm are described. One type works in the geometrical optics regime and applies to the testing of ground-based optical telescopes through the atmosphere. The other type works in the diffraction regime and applies to optical telescopes in space or ground-based telescopes operating in the thermal infra-red and beyond.

1. Introduction

The development of large telescopes both in space and on excellent ground sites is putting a high demand on the performance of telescope optics. Examples are the Hubble Space Telescope and the Canada-France-Hawaii Telescope under good seeing conditions. In both cases image quality is limited by the telescope optics. In order to control telescope optical quality on a permanent basis there is a definite need for new, simpler but still accurate, optical testing methods. It is shown here that the optical quality of a telescope can be quantitatively analyzed by simply recording a small set of properly defocused images with a CCD camera of good photometric quality. Image processing algorithms have been developed to obtain accurate estimates of the aberration terms as well as complete mirror figure errors. Two types of algorithms have been studied.

One type works in the geometrical optics regime and applies to the testing of ground-based optical telescopes through the atmosphere. In this case extra-focal images are recorded outside the so-called caustic zone, far enough from the focal plane for diffraction effects to be negligible. In this regime the intensity variations over the image surface essentially reflect changes in the wave-front total curvature

---

[*]University of Hawaii, Institute for Astronomy, 2680 Woodlawn Drive, Honolulu, HI 96822, USA.

(Laplacian) and are used to map these changes. For this reason the method was originally called wave-front curvature sensing in contrast with current wave-front slope sensing methods. The method shares many common properties with incoherent wave-front sensing techniques such as the Hartmann sensing method. Both techniques can be used with broad-band light and extended sources such as turbulence-broadened stellar images with similar sensitivities. However, the extra-focal image technique is easier to implement and to calibrate.

The other type of algorithm works in the diffraction regime and applies to optical telescopes in space or ground-based telescopes operating in the infra-red and beyond. In this regime extra-focal images are recorded well within the caustic zone and diffraction effects dominate. Iterative phase retrieval algorithms are used to reconstruct the wave-front surface from such images. This wave-front reconstruction method shares the same limitation as coherent, interferometric sensors. Its use is limited to point sources and quasi-monochromatic light.

2. The curvature sensing technique

The technique [Roddier, 1988], [Roddier *et al.*, 1988] consists of recording highly defocused stellar images on a CCD camera without any additional optics. Exposure times of a few tens of seconds are used to average out seeing effects. A pair of images symmetrically defocused on each side of the focal plan is required. At the small perturbation approximation, the relative difference in the irradiance distribution is given by

$$\frac{I_1 - I_2}{I_1 + I_2} = \frac{f(f - l)}{l}\left[\frac{\delta W}{\delta n}\delta_c - P\nabla^2 W\right]$$

where I_1 and I_2 are the illuminations recorded in the two images, f is the focal length and l the distance to focus. The expression between brackets contains two terms. The first term called the edge signal is an impulse distribution δ_c produced by local shifts at the beam edge. It is a measure of the wave-front edge slope $\delta W/\delta n$ in a direction perpendicular to the edge. The second term called the curvature signal is a measure of the local wave-front Laplacian $\nabla^2 W$. P is the pupil transmission function assumed to be one inside the pupil and zero outside. The wave-front is reconstructed by solving a Poisson equation using the edge signal as a Neumann boundary condition [Roddier, 1991]. A new algorithm based on iterative Fourier transforms has been developed for this purpose. It can be used to reconstruct wave-fronts either from slopes or from Laplacians [Roddier & Roddier, 1991].

In practice the edge signal has a finite width which is a measure of the spatial resolution on the reconstructed wave-front. In case of large aberrations the difference between the two defocused images increases and the spatial resolution decreases. To maintain the spatial resolution one can decrease the difference between the two images by increasing the distance to focus. This is done at the expense of a lower sensitivity, thus setting a limit to the dynamic range of the method. As described,

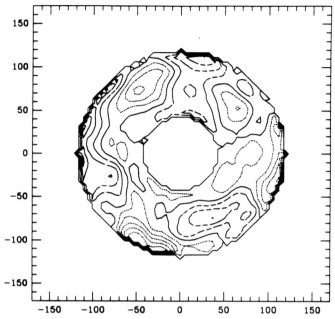

Figure 1. Primary mirror figure of the UH 88-in telescope reconstructed from out-of-focus images.

it is comparable to that of a Hartmann test with quadrant detectors. However, the effect of a large aberration is to produce a theoretically known image distortion. Starting with an estimate of the aberration one can remove its effect on the image by calculation, and thus improve the dynamic range of the wave-front reconstruction method. For instance the image overlap can be improved by translating the images to remove any residual tilt error, or by changing their magnification to match their size better, which is a means to remove defocus terms. Similar but more elaborate transformations can be used to remove higher order terms such as coma or spherical aberration [Roddier & Roddier, 1991; Roddier et al., 1990].

The method has been successfully used to test ground-based telescopes on Mauna Kea. Fig. 1 shows the primary mirror figure of the UH 88" telescope reconstructed from our-of-focus images recorded with a CCD camera at prime focus. The effect of the strong spherical aberration produced at prime focus by this Ritchey–Chretien telescope was removed from the out-of-focus images by calculation, before subtracting oppositely defocused images. The wave-front was then reconstructed by solving the Poisson equation. Fig. 1 shows a contour plot of the residual mirror surface errors after removal of tilt, defocus, coma, astigmatism and spherical aberration. The contour line interval is 25nm. The dotted contour shows positive values. The mirror surface is quite smooth. The largest errors at the edge are related to known defects in the mirror blank. From the observed amount of spherical

Table 1. Observed rms spherical aberration

Telescope	Spherical aberration (μm rms)
UH 88-in	0.06
CFHT	0.19
IRTF	0.35

aberration, the conical constant of the primary was estimated to be -1.0535 ± 0.005 instead of the expected value -1.0497. A new secondary is being made to cancel the error. Astigmatism which is removed in Fig. 1 was found to be essentially produced by mirror support problems. A similar situation was found for all the telescopes we have tested (CFHT, IRTF). Primary mirrors are quite smooth but poorly supported. The two main causes of image degradation were found to be astigmatism due to mirror support and spherical aberration due to inaccurate conical constants. Table 1 shows the rms spherical aberration measured on three telescopes on Mauna Kea.

3. Phase retrieval techniques

Wave-front reconstruction from defocused images can be extended to the diffraction regime by means of phase retrieval algorithms. We have recently applied such algorithms to defocused stellar images recorded with the Planetary Camera (PC) on board of the Hubble Space Telescope (HST), providing a means to test the telescope in flight. The exact pupil geometry was found to be easily recovered after a few iterations using a Gerchberg–Saxton algorithm [Gerchberg & Saxton, 1972] with loose pupil-plane constraints. Basically one starts with a first wave-front estimate and a uniformly illuminated pupil and compute the expected amplitude and phase in the image plane. The amplitude is replaced by the observed amplitude (square root of the illumination), and an inverse transform is taken producing a new estimate of the wave-front and of the pupil illumination. The telescope spider arms and central obstruction usually appear after a single iteration and can be sharpened by refining tilts and defocus error terms. Fig. 2 shows an example of defocused image recorded by HST. Fig. 3 shows the pupil illumination reconstructed from such an image after a few iterations. One can see both the telescope and the PC secondary mirror supports. They are not aligned. We have determined that they become aligned when the star image coordinates are at pixel 245,310, instead of being at the camera center at pixel 400,400 showing evidence for a camera misalignment.

Once the dark areas in the pupil plane are accurately located, their amplitude is set to zero and the dominant aberration terms (in this case defocus and spherical aberration) are refined to minimize the difference between the observed and the computed image. Up to 55 additional Gerchberg–Saxton iterations were used to

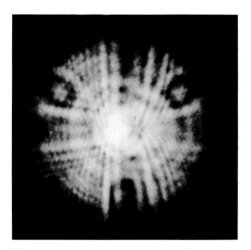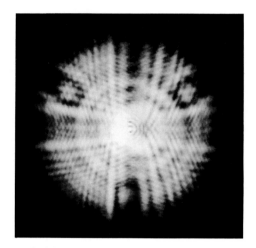

Figure 2. Left: example of defocused image recorded by the Hubble Space Telescope. Right: same image computed from the reconstructed wave front.

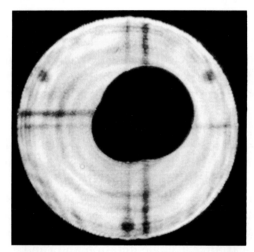

Figure 3. Reconstructed HST pupil.

refine our wave-front estimate further. Fig. 2 shows a comparison between an observed image and the same image computed from the estimated wave-front and pupil geometry. Compared to the computed image, the observed image generally appears to be slightly blurred by telescope jitter during the exposure. This effect limited the accuracy of our results. At the PC focal plane the spherical aberration was estimated to be -0.290μm rms with an uncertainty of ±5nm. A gray level map of the residual wave-front errors, that is after removal of tilts, defocus and spherical aberration, is shown in Fig. 4. It shows circular groves which are consistent

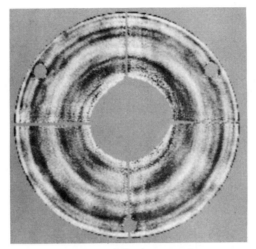

Figure 4. HST residual wave-front errors.

with known groves produced by the polishing tool both on the primary and the secondary. The rms amplitude of these groves is 10 nm on the mirror surface.

The same phase retrieval algorithm has recently been applied to narrow-band short exposure infra-red images recorded at 4μm with an IR detector array using the Infra-Red Telescope Facility (IRTF) on Mauna Kea. At this wavelength the effect of atmospheric turbulence is essentially image motion and is eliminated in a short exposure. Permanent telescope aberrations can be retrieved by averaging a few tens of reconstructed wave-fronts. The reconstructed wave-front is shown in Fig. 5. It is dominated by spherical aberration and decentering coma.

4.   Conclusion

We have described new algorithms developed at the University of Hawaii to test astronomical telescopes from out-of-focus images.

One method is based on the geometrical optics approximation. It consists of taking the difference between the illuminations in two symmetrically defocused stellar images. For sufficiently large defocus values this difference maps the wave-front local curvature (Laplacian) and the wave-front radial edge slope. The wave-front is reconstructed by solving a Poisson equation with the estimated edge slope as a Neumann-type boundary condition. The method is well suited to test ground-based telescopes on stellar sources at visible wave-lengths. Like the classical Hartmann test, it works with broad-band light, and in spite of the light spread produced by atmospheric turbulence. In addition it is easier to implement and to calibrate. It has been successfully used to test telescopes on Mauna Kea and is now being used at other observatories.

The other method we described works in the diffraction regime and requires

241

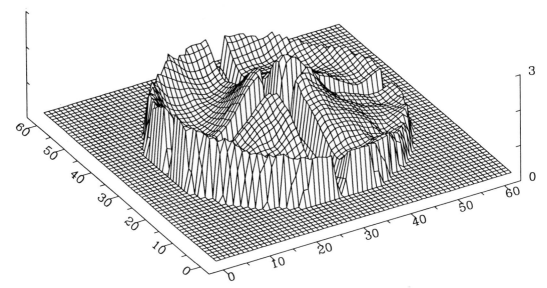

Figure 5. Wave front reconstructed from IRTF Cassegrain focus data taken at 4 μm. The vertical scale is in μm.

narrow-band light. It is more sensitive to the effect of atmospheric turbulence and is best suited to test ground-based telescopes at longer wavelengths (infra-red or millimetric) or telescopes in space at shorter wave-lengths. It is based on iterative phase retrieval algorithms. It has been successfully used to estimate the aberrations of the Hubble Space Telescope. Not only the wave-front surface but also the illumination in the pupil plane was reconstructed showing evidence for a camera misalignment. The technique has also been recently used to test the Infra-Red Telescope Facility on Mauna Kea.

Acknowledgments

This research was supported partly on University of Hawaii funds, partly on a grant from the Strategic Defense Initiative Organization, Office of Innovative Science and Technology, and managed by the Harry Diamond Laboratories. The data from the Space Telescope were processed under contract with the Jet Propulsion Laboratory.

References

[Gerchberg & Saxton, 1972] Gerchberg R.W., Saxton W.O., *A practical algorithm for the determination of phase from image and diffraction plane pictures. Optik, Vol. 35 (1972), p. 237.*

[Roddier, 1988] Roddier F., *Curvature sensing and compensation: a new concept in adaptive optics. Appl. Opt., Vol. 27 (1988), p. 1223.*

[Roddier, 1991] Roddier N., *Algorithms for wave-front reconstruction out of curvature sensing data. Proc. SPIE, Vol. 1542 (1991), p. 120.*

[Roddier *et al.*, 1988] Roddier F., Roddier C., Roddier N., *Curvature sensing: a new wave-front sensing method. Proc. SPIE, Vol. 976 (1988), p. 203.*

[Roddier *et al.*, 1990] Roddier C., Roddier F., Stockton A., Pickles A., Roddier N., *Testing of telescope optics: a new approach. Proc. SPIE, Vol. 1236 (1990), p. 756.*

[Roddier & Roddier, 1991] Roddier F., Roddier C., *Wave-front reconstruction using iterative Fourier transforms. Appl. Opt., Vol. 30 (1991), p. 1325.*

20
Super seeing from Antarctica?

Peter Gillingham [*]

Abstract

Potential gains for optical/infrared observation from Antarctica are discussed, together with the programme to determine the reality of such possibilities.

1. Introduction

Astronomers have become more and more aware in recent years that Antarctica has much to offer in observing conditions. Obvious benefits arise directly from proximity to the geographic pole, e.g. capabilities for continuous observations for periods of 100 hours or more; others follow from the extremely dry, tenuous, and cold atmosphere. In winter, near the highest part of the plateau (which is more than 4000 m above sea level) precipitable water vapour should often be below 0.1 mm, the pressure is the equivalent of an altitude above 5000 m, and temperatures fall below −80°C. These conditions greatly favour infrared and mm-wavelength observations, because of good atmospheric transmission and low background from the atmosphere and the telescope. Another important attraction at a very high altitude Antarctic site may be better seeing than from anywhere else on the Earth's surface.

2. Considerations and observations

At most observatory sites, diurnal temperature variation is a major cause of seeing degradation, both in the free atmosphere and, especially for large instruments, in the vicinity of the telescope. In Antarctica, at least during the dark months, systematic diurnal variation in temperature is negligible, so seeing will not suffer on this account. The Antarctic atmospheric circulation pattern centres on the highest point of the plateau, where the predominant airflow is a slow settling to feed the very consistent downward-flowing surface winds. This should promote extraordinarily stable and uniform optical quality in the atmosphere above the highest point.

There have been few measurements of seeing reported from high-altitude Antarctic sites – just some visual observations from the South Pole, which is at about 2800 m

---

[*]Anglo-Australian Observatory, P.O. Box 296, Epping, NSW, Australia 2121.

(e.g. Harvey [1989, 1991]). These have not been remarkably good but most have been in daytime (i.e. summer) and have been made from small telescopes a few metres at most above the snow. Near-ground effects can dominate seeing degradation at many observing sites – hence the elevation of telescopes on piers and towers – but this is especially likely in Antarctica. There the ground cools well below the bulk air temperature, setting up a strong inverse air temperature gradient in the lower several to many tens of metres. In winter, the gradient can reach $0.1°C/m$ or more. Stirring of this boundary layer leads to quite large air-temperature variations near the surface from hour to hour and, presumably, to significant micro-thermal fluctuations which will degrade seeing.

Crucial uncertainties which must be resolved before we will know whether 'super seeing' is economically attainable from Antarctica are the quality of seeing from above the boundary layer and the height to which a telescope would have to be raised to avoid significant seeing degradation by this layer. Part of the programme of CARA (the US Center for Astrophysical Research in Antarctica) is to measure seeing at the South Pole more quantitatively than in the past. It is important (a) that this programme be extended to include measurements which will resolve the uncertainties just mentioned and (b) that, as soon as appropriate techniques for doing this are proved at the Pole, the same measurements be made at higher altitude sites. Initial moves have been made towards setting up a collaboration to this end, involving CARA, the author, and a group at the University of Nice, who have very relevant experience in site testing for seeing (e.g. [Vernin & Azouit, 1990]) as well as experience in building optical instruments for operation at the South Pole (for solar seismology). Encouraging contact has also been made with Soviet astronomers, which might lead to seeing tests being made at the Soviet Antarctic station, Vostok, currently the highest manned inland base.

3.   Conclusions

It is possible that an Antarctic site, particularly one at the highest part of the plateau, will, in addition to providing great gains over temperate sites in other respects, allow observations with seeing unprecedented from the ground.   The longer-wavelength part of the K atmospheric window, centred around 2.4 μm, is free of atmospheric emission lines and the thermal radiance from the atmosphere is very steeply dependent on temperature, such that Antarctic sites offer about a 100-fold reduction compared with sites at $0°C$. At 2.4 μm, the diffraction-limited resolution (the Airy disc radius) of a 3-m telescope is 0.2 arcsec. So with a telescope of this size or larger, seeing which frequently allowed near diffraction-limited performance would further enhance the Antarctic sensitivity gain and deliver very high resolution indeed. There is reason to hope such seeing will be attainable at a very high Antarctic site but it may be necessary to mount the telescope higher above the terrain than is usual on a mountain-top site.

References

[Harvey, 1989] Harvey J., *Solar observing conditions at the South Pole. in AIP Conference Proceedings 198, "Astrophysics in Antarctica", Eds Mullan, Pomerantz, & Stanev, 1989.*

[Harvey, 1991] Harvey J., *Daytime astronomical observing conditions at the South Pole. in Joint Commission Meeting,"The Development of Antarctic Astronomy", IAU General Assembly, Buenos Aires 1991. Proc. IAU General Assembly (1992), in press.*

[Vernin & Azouit, 1990] Vernin J., Azouit M., *SCIDAR measurements during LASS-CA. in ESO Report No. 60, "Site Testing for the VLT, the La Silla Seeing Campaign" Ed. Sarazin M., 1990.*

21
Science with HRCam at the CFHT

René Racine and *Robert D. McClure* [†]

Abstract

We briefly review the scientific motivations which led to the construction of the DAO/CFHT "High Resolution Camera" (HRCam) and the main characteristics of the instrument. Three years of experience at the telescope have revealed which features of the instrument are appreciated by the users, which are detested and which are indifferent. The broad spectrum of research being carried out with HRCam is summarized and illustrated by images of sub-half-arcsec quality.

1. Introduction

We study star clusters and stellar populations in galaxies. Crowding and background noise impose the ultimate observational limits to our work. Aware, as are all astronomers, that shrinking the PSF is directly equivalent to increasing the telescope aperture – with the added benefits of better resolution and reduced crowding – "image quality" (IQ) is of primordial importance to us; the slightest gain is perceived as an enormous advantage. Access to the Canada-France-Hawaii Telescope (CFHT) has made us acutely aware of the benefit of the "good seeing" frequently provided by that instrument and which facilitates otherwise impossible research projects. At the same time, our appetite for the even better IQ, which appeared within reach, has been irresistibly wetted. Early experiments at the CFHT [Christian, 1983] had shown that if a practical imager was available which would freeze the fast image motion so apparent at times of excellent "seeing" and which could select those instants of best IQ, superior observations would be obtained. The idea was not new, of course [Leighton, 1956] and Laird Thompson's Image Stabilizing Instrument System (ISIS) [Thompson & Ryerson, 1983] had already pioneered some of the techniques on Mauna Kea. In August 1986, during IAU Symposium 126 on globular clusters celebrating the 100th anniversary of the birth of Harlow Shapley, we convinced

[*]Département de Physique, Université de Montréal, and Observatoire du Mont Mégantic, Montréal, Québec, Canada H3C 3J7.
[†]Dominion Astrophysical Observatory, National Research Council, Victoria, Canada V8X 4M6.

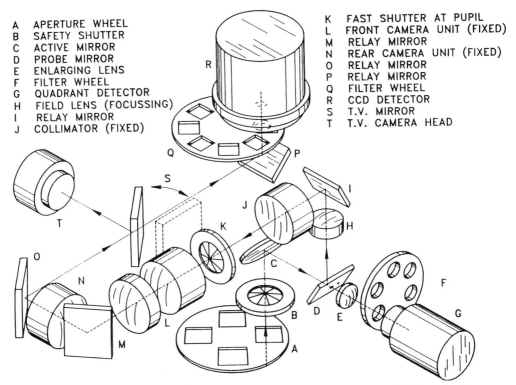

A   APERTURE WHEEL
B   SAFETY SHUTTER
C   ACTIVE MIRROR
D   PROBE MIRROR
E   ENLARGING LENS
F   FILTER WHEEL
G   QUADRANT DETECTOR
H   FIELD LENS (FOCUSSING)
I   RELAY MIRROR
J   COLLIMATOR (FIXED)

K   FAST SHUTTER AT PUPIL
L   FRONT CAMERA UNIT (FIXED)
M   RELAY MIRROR
N   REAR CAMERA UNIT (FIXED)
O   RELAY MIRROR
P   RELAY MIRROR
Q   FILTER WHEEL
R   CCD DETECTOR
S   T.V. MIRROR
T   T.V. CAMERA HEAD

Figure 1. Schematic layout of HRCam optics. See text for a brief description.

ourselves that the time was ripe and the opportunity at hand to design and build HRCam, as a collaborative project between the DAO and the CFHT. In June 1988, HRCam was doing science at the CFHT.

2.   The HRCam concept

The most important characteristics of the instrument had to be user-friendliness and reliability of operation. We felt that our colleagues, and indeed ourselves, would have little desire to commit precious CFHT time to an instrument likely to reduce efficiency in telescope use or even to lose nights to breakdowns. At the same time, various modes of operation had to be provided to allow trade-offs between resolution and other parameters – such as speed – depending on the nature of the program. We also envisaged HRCam as a practical first step toward more sophisticated adaptive-optics imagers for which experience at the telescope had to be gained. Thus, HRCam is little more than a "fast guider" whose goal is to remove wavefront tilt only, whether due to atmospheric turbulence or, more prosaically, to tracking errors, wind buffeting, etc.

The theoretical ideas on which we based our expectation of performance were

Figure 2. The DAO/CFHT HRCam fully assembled. The multitude of optional functions initially provided made for a rather crowded opto-mechanical environment.

most naive, as we were to later discover. They were basically limited to Fried's [1966] demonstration of how the PSF spread angle can be reduced by removing the tilt component of the wavefront aberration due to the atmosphere. By good CFHT "seeing" (0.5 arcsec FWHM, r_o = 25 cm at 650 nm), we expected, from theory, a gain of 2 × in resolution by using a ~1 m sub-aperture (D/r_o = 4) and of ~ 1.2 × at full aperture. Even that lower gain at 3.6-m aperture appeared most desirable. And then further gain could be had by image selection! HRCam was thus designed as a "friendly" fast-guiding CCD imager, great attention being paid to the ability to operate in the sub-aperture and/or image-selection modes while preserving the photometric integrity of the field.

It is important to mention that while this HRCam concept of real-time stabilization and selection for an integrating (CCD) imager was being developed, a parallel effort was under way at the CFHT to develop the Segmented Pupil Imager (SPI) [Lelièvre et al., 1988], a photon-counting system intended for post-detection selection, centroiding and co-adding of individual short exposures. The limited dynamic range and field of view of the then available photon-counting camera imposed rather narrow limits on the magnitude of suitable reference sources ("guide stars") and limited the SPI to carefully selected science fields. We also viewed HRCam's ability to provide a final integrated image at the telescope for evaluation as advantageous

Table 1. HRCam Observing Programs, July 1988 – December 1991

	Short program title	No. of runs	No. of nights
1	Imaging of Planets	1	1
2	Optical Jets in Young Stars	2	4
3	Protostellar Objects	2	3
4	Dust in Circumstellar Envelopes	1	3
5	Protoplanetary Nebulae	2	5
6	Mass Segregation in M13	3	8
7	The Core of M15	2	2
8	Globulars Near the Galactic Center	1	2
9	Ages of Outer Globulars	1	3
10	HB Stars in Fornax Globulars	1	2
11	Structure of M31 Globulars	1	2
12	CMDs of M31 Globulars	3	7
13	Cores of Early Type Galaxies	4	8
14	Cores of Dwarf Ellipticals	3	8
15	Photometry of Black Hole Candidates	2	5
16	CMDs for Magellanic Dwarfs	1	2
17	Surface Photometry of Ellipticals	1	4
18	Globulars around cD Galaxies	1	4
19	Post-Starburst Galaxies	3	6
20	Structure of Cygnus A	1	2
21	Imaging of Radio Galaxies	1	1
22	Variable Stars in Virgo Spirals	2	1
23	H_0 from Old Stellar Populations	1	4
24	Distances from Surface Bright. Fluct.	2	6
25	Imaging of Bright QSOs	3	6
26	Gravitational Lenses among QSOs	2	6
27	Galaxies Associated with Quasars	1	2
		48	107

for the management of the night work. Real-time corrections do require the integration into the instrument of more complex opto-mechanical and control systems, but post-detection corrections must provide these later in software development and computing time. The added complexity of a real-time instrument is a necessary price to pay for the efficient use of the telescope time.

Fig. 1 shows schematically how the various components are arranged in HRCam

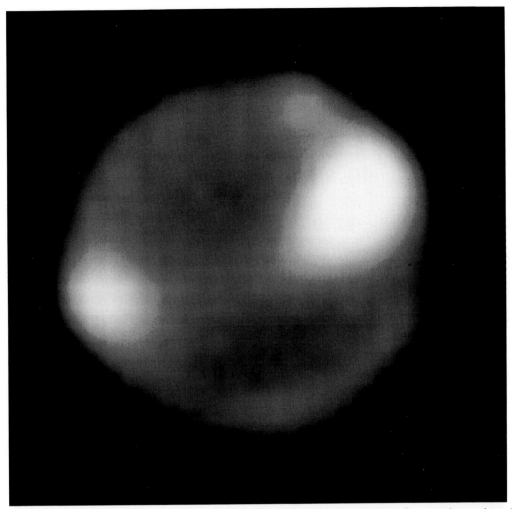

Figure 3. An image of the planet Neptune taken in the light of a methane band. The angular diameter of the disk is 2 arcsec. This smoothed image clearly shows two active regions and a noticeable limb brightening. The resolution (FWHM ~ 0.4 arcsec) is sufficient for global synoptic studies of the planet's climate.

and Fig. 2 shows what a maze they in fact produce. A detailed description of the instrument is given by McClure *et al.* [1989]. The "heart" of the instrument is the piezo-activated tip-tilt mirror which folds the beam before the telescope prime focus. An inclined field mirror is located at that focus and a small aperture, 4 arcsec in diameter, lets the light from the guide star pass to a quadrant detector located behind this "probe" mirror. The imbalance of the flux on the elements of the quadrant detector is the error signal which is made to drive the piezo-actuators and the guiding

251

Figure 4. The planet Pluto and its satellite Charon at a separation of 0.7 arcsec (FWHM = 0.52 arcsec).

mirror. The closed-loop bandpass of the system is ~ 20 Hz. Transfer optics correct the coma of the parabolic primary mirror, produce a pupil image where masks and a fast shutter can be located, and re-image the field with a magnification of 2 on the CCD where a 15-μm pixel corresponds to 0.10 arcsec.

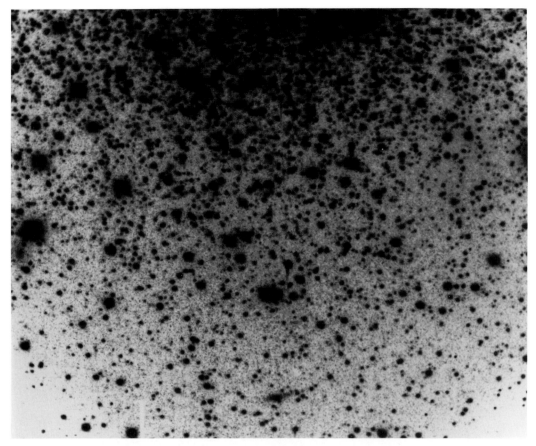

Figure 5. A region in the globular cluster NGC 6838 (FWHM = 0.48 arcsec). High resolution allows study of stars in the inner regions of these clusters. This reveals how objects of different masses (luminosity) have been dispersed by the dynamical evolution of the clusters.

3.   The HRCam experience

HRCam's user-friendliness has made it a popular and successful instrument. Although it is rather complex – unnecessarily so, as is now realized (see below) – its control software and user interface are such that observers feel confident in its use after only a few exposures. Over the last three years, demand and use have made HRCam one of the most popular and most frequently scheduled of all intruments at the CFHT. Some 25 HRCam science papers have appeared in print and at least a dozen more have been accepted for publication. Applications have ranged from planetary astronomy to extragalactic and cosmological researches.

Yet the gain in resolution achieved by HRCam over the conventional FOCAM imager is quite modest. Statistics show an average IQ of $\langle \text{FWHM} \rangle = 0.56$ arcsec with HRCam versus 0.75 arcsec with FOCAM, a mere 1.3 × improvement. But for

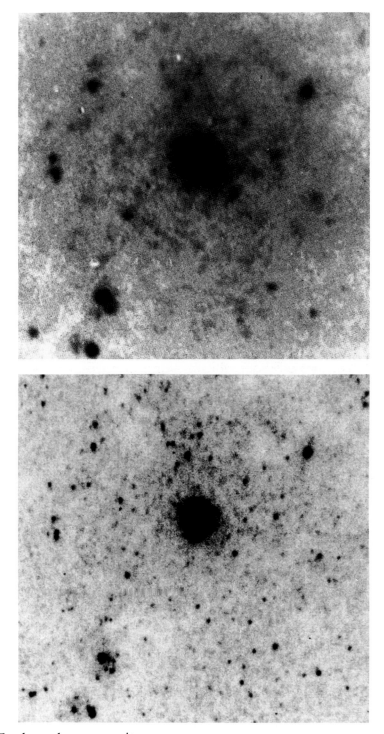

Figure 6. For legend see opposite

many programs on faint crowded stars, where the S/N ratio goes as D/ω^2 (Racine, this volume), the performance of the 3.6-m CFHT with HRCam is equivalent to that of a 6-m telescope on Mauna Kea with FOCAM. And this is sufficient to make HRCam an attractive option! More important perhaps is the fact that HRCam achieves sub-half-arcsec FWHM 10 times more frequently than FOCAM and allows otherwise impossible research to be carried out.

The main limitations of the instrument are its small field of view (2 arcmin diameter), designed as it was with, at most, 1024^2 CCDs in mind (how conservative!) and for 0.10 arcsec/pixel sampling, and the ensuing requirement to find a suitably bright guide star within 2 arcmin of the science target. This last limitation has been much reduced in 1991 when the quadrant-detector photomultipliers were replaced by higher-efficiency and broader-bandpass avalanche photodiodes, thereby gaining 2 mag in sensitivity. Adequate guiding (10 Hz) at full aperture is now possible on a star of $V < 19.5$ mag and, even at the galactic poles, most fields provide a good selection of reference stars.

The main feature which made HRCam unnecessarily complicated is the requirement to produce a sharp pupil image where the fast shutter (for image selection) was to be located and where pupil masks (for reduced D/r_o) can be placed. This made the design of the re-imaging optics complex and the opto-mechanical construction very crowded. It also required developing a mechanical shutter assembly capable of $> 10^4$ actions per hour with < 10 msec response time (and one which can be easily replaced in case of "melt-down"!), an image quality evaluator via a 1.2-arcsec diameter guide hole, and the associated electronics and control software. None of this has attracted much sympathy from the users. We do not even know yet if the image selector works properly! The plain fact is that astronomers are extremely reluctant to cut down the mirror surface by a factor of 10 and/or leave the shutter closed for $> 90\%$ of the time – which would be required for selection to really make a difference. And Time Allocations Committees are no less reluctant to grant observing time to astronomers who propose to do some or all of the above! The lesson clearly is that if a high-resolution imager – or any instrument – is to perform better with a reduced telescope aperture, it should be installed on a smaller telescope!

Another feature which causes much grief to the observers is the unavoidable collection of microscopic dust on the in-focus probe mirror and whose images pepper the CCD field. Since the mirror can be moved to optimize the positions of the guide star and science target on the CCD, this can make the acquisition and

---

Figure 6. These pictures show the nucleus of the bright spiral M33 in red light. The upper frame, taken with a conventional camera, has FWHM = 0.8 arcsec. The lower frame, taken with HRCam, has an image quality of 0.35 arcsec. Studies based on frames such as this one, combined with spectroscopy, show that the centre of M33 does not contain a massive black hole [Kormendy & McClure, 1992].

Figure 7. The globular star cluster G11 in the Andromeda galaxy. Resolution into stars (FWHM = 0.46 arcsec) allows the colour–magnitude diagrams of such clusters to be studied down to the horizontal branch level at V ~ 25 mag.

application of flat-field frames a real pain! Consequently, users have found it very desirable, and quite satisfactory, to adopt a fixed guide-position at the edge of the field, and rotate the instrument to acquire a guide star. Two lessons here: (1) no in-focus optical surfaces please!; (2) reduced flexibility can be a virtue.

But a number of HRCam features are much liked. The user interface provides X–Y displays of the tip-tilt mirror position and of the error signal. This proved essential to monitor in real time the performance of the system and to optimize parameters such as sampling time and servo gains. The faintly-reflecting guide hole, etched into the protected silver coating of the probe mirror substrate, makes a very

Figure 8. The spiral galaxy NGC 4571 in the Virgo cluster is clearly resolved into stars in this 6000-second exposure (FWHM = 0.42 arcsec) reaching an R magnitude of 26. These images are being used to study the light curves of long-period variables and, hopefully, Cepheids in Virgo spirals. This should give a precise distance to the Virgo cluster, a fundamental yardstick in observational cosmology [Pierce *et al.*, 1992].

convenient mask for the study of faint "fuzz" around quasars and faint companions to bright stars. A feature which was implemented with some urgency after the first runs is a link to the telescope control system. This enables the TCS to remove the slow large-amplitude tracking errors – the HRCam optics do not allow the use of the conventional autoguider – removing the need to hand-guide from the X–Y position display, a most painful occupation when a brief mistake of ~ 1 arcsec would cause HRCam to loose lock and spoil a long exposure.

We now know that if HRCam achieves the resolution gain we observe, and expected, of 1.3 × at full aperture, and of ~ 2 × with a 1.2-m diameter subpupil

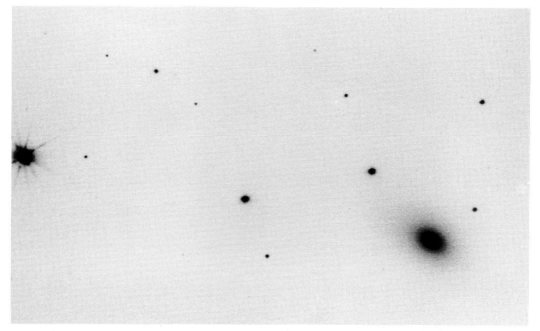

Figure 9. This frame of the field around the elliptical radio-galaxy NGC 7052 was taken with a 1.2-m sub-pupil and is one of the best in resolution obtained with HRCam. The bright guide star in its faintly reflecting "hole" is conspicuous at the left of the field. Its FWHM is 0.24 arcsec. The increasing elongation of the stellar images as the distance from the guide star increases is due to the anisotropy of the isoplanatism [McClure *et al.*, 1991]. At the position of the diffuse galaxy image (lower right), image quality has deteriorated to 0.32 arcsec.

where 0.3 arcsec FWHMs are obtained, it does so, in good part, "for the wrong reasons", i.e. not exclusively by correcting the tilt component of the wavefront aberrations introduced by atmospheric turbulence. It also corrects residual telescope tracking errors and wind buffetting, which the conventional auto-guider fails to do perfectly. With a reduced aperture, one mostly reduces the importance of the slight residual figuring and collimation errors in the optics. This results in higher resolution at times of good "seeing". To be sure, "real" atmospheric corrections are also produced, as anisoplanatism across some images demonstrates, but they may not be as beneficial as our understanding of the theory would lead us to believe. Undoubtedly, the performance of HRCam is not perfect and the theoretical limit is not reached. But the possibility remains that the operational implementation of even the simplest adaptive optics principles requires considerable experimentation before a near-ideal approach is discovered. For instance, a quadrant detector, which in principle measures the median of the light distribution in an image, may not be as good an operational choice for centroiding as another detector which would measure other statistics of the image intensity.

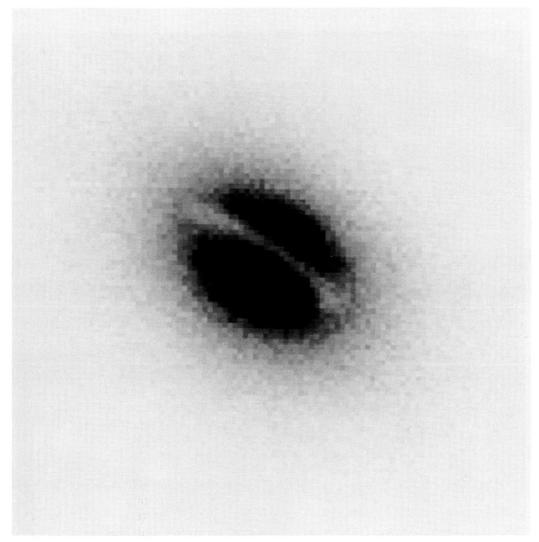

Figure 10. The core of NGC 7052 shows a ring of dust less than 0.2 arcsec across in orbit around the nucleus of this elliptical galaxy. Since dust is not normally present in ellipticals, the ring may result from NGC 7052 cannibalising a (dusty) spiral galaxy it encountered in space [Nieto *et al.*, 1990].

4. The HRCam science

Table 1 gives a list of the HRCam projects scheduled on the telescope since the commissioning of the instrument in June 1988, and the total number of nights allocated to each. In some cases where very similar projects are carried out by different teams, a single entry is made and the nights added.

Generally, the research topics for which observers request HRCam time are not

Figure 11. Montage from an image of the faint galaxy 2237+0305 (FWHM = 0.42 arcsec) showing (large frame) the outline of the faint disk and the bright central bulge and bar. The small central insert, printed at reduced intensity, shows the nucleus of the galaxy to have a peculiar cross-like appearance. An enlargement (lower right) reveals four stellar images within a 1.8 arcsec diameter circle around the nucleus. Spectra have shown these to be four images of the same distant quasar seen through the foreground galaxy: a mirage due to the bending of the quasar light by the gravitational field of the galaxy. The upper left insert (1.0 arcsec resolution) can be compared to the full frame to appreciate the gain better IQ provides [Racine, 1991].

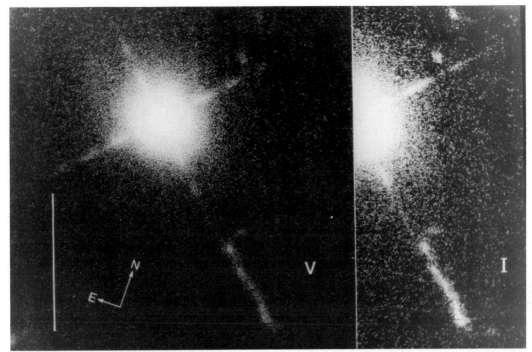

Figure 12. The bright quasar 3C273 and its jet in visual and infrared light. The knotty structure of the jet is apparent [Hutchings & Neff, 1991].

qualitatively different from those they would address with a conventional imager at CFHT or elsewhere. We take this as an indication of HRCam success as a research instrument: it is used as a versatile – and superior – tool in a diversity of programs and does not itself dictate the research. Of course, HRCam projects tend to be quantitatively more demanding in terms of resolution or limiting magnitude. Many remain high-risk projects as they often demand the ultimate performance HRCam is able to deliver under the best "seeing" conditions. For that reason, the TAC wisely insists that competitive back-up projects be presented by those prospective HRCam users who propose programs at the very edge of feasibility.

Finally, we illustrate the scientific performances of HRCam with a small collection of frames (Figs. 3–13). Except for some interpolative smoothing for display purposes where indicated, no deconvolution or image sharpening was applied to any of these 0.13 arcsec/pixel frames whose FWHMs range from 0.24 to 0.52 arcsec.

5. Conclusion

HRCam remains a modest step toward improved-resolution imaging in astronomy, but one from which much has been learned about the practical scientific use of such devices. In addition, HRCam has helped gain a more quantitative understanding

261

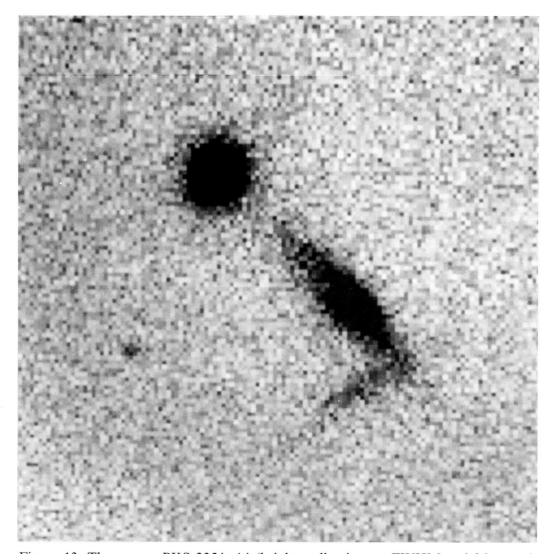

Figure 13. The quasar PKS 2354+14 (bright stellar image, FWHM = 0.36 arcsec) may be associated with or gravitationally-lensed by one of two faint lenticular galaxies [Crampton *et al.*, 1989].

of the various factors which affect image quality at the CFHT [Racine *et al.*, 1991]. With help from the natural qualities of the Mauna Kea site, HRCam has accustomed the CFHT community to think in terms of and plan for half-arcsec image quality. And we believe that when instrumental improvements in progress are fully implemented (Racine, this volume), the criterion will approach 0.3 arcsec. Science with HRCam at the CFHT demonstrates rather vividly the enormous benefit even a modest improvement in optics has on astronomy. The authors, and their DAO and CFHT colleagues, expect Professor Wynne will approve!

References

[Christian, 1983] Christian C.A., *Proc. SPIE, Vol. 445 (1983), p. 484.*

[Crampton *et al.*, 1989] Crampton D., McClure R.D., Fletcher J.M., Hutchings J.B., *Astron. J., Vol. 98 (1989), p. 1188.*

[Fried, 1966] Fried D.L., *J. Opt. Soc. Amer., Vol. 56 (1966), p. 1380.*

[Hutchings & Neff, 1991] Hutchings J., Neff S.G., *Publ. Astron. Soc. Pacif., Vol. 103 (1991), p. 26.*

[Leighton, 1956] Leighton R., *reported in Amateur Scientist, Scientific American, Vol. 194 (1956), p. 156.*

[Lelièvre *et al.*, 1988] Lelièvre G., Nieto J.-L., Salmon D., Llebaria A., Thouvenot E., Boulesteix J., Le Coarer E., Arnaud J., *Astron. Astrophys., Vol. 200 (1988), p. 301.*

[Kormendy & McClure, 1992] Kormendy J., McClure R.D., *Astron. J., (1992), in press.*

[McClure *et al.*, 1989] McClure R.D., Grundmann W.A., Rambold W.N., Fletcher J.M., Richardson E.H., Stillburn J.R., Racine R., Christian C.A., Waddell P., *Publ. Astron. Soc. Pacif., Vol. 101 (1989), p. 1156.*

[McClure *et al.*, 1991] McClure R.D., Arnaud J., Fletcher J.M., Nieto J.-L., Racine R., *Publ. Astron. Soc. Pacif., Vol. 103 (1991), p. 570.*

[Nieto *et al.*, 1990] Nieto J.-L., McClure R., Fletcher J.M., Arnaud J., Bacon R., Bender R., Comte G., Poulain P., *Astron. Astrophys., Vol. 235 (1990), p. L17.*

[Pierce *et al.*, 1992] Pierce M., *et al., Astrophys. J., (1992), in press.*

[Racine, 1991] Racine R., *Astron. J., Vol. 102 (1991), p. 454.*

[Racine *et al.*, 1991] Racine R., Salmon D., Cowley D., Sovka J., *Publ. Astron. Soc. Pacif., Vol. 103 (1991), p. 1020.*

[Thompson & Ryerson, 1983] Thompson L., Ryerson H., *Proc. SPIE, Vol. 445 (1983), p. 503.*

V Reflections

22

Grubb Parsons – from War to Peace

G.M. Sisson *

Abstract

George Sisson, OBE, joined Grubb Parsons in 1940 as Chief Engineer; he became General Manager in 1945, non-executive Chairman in 1969, and retired in 1974. What follows are some of his personal recollections through the period in which Grubb Parsons telescopes were increasingly influential in UK and international astronomy — *Ed.*

1. Introduction

The history of Thomas Grubb and his son Howard engaged in making telescopes at Rathmines near Dublin during the last century has been well documented. Suffice it here to say that in the First World War the business of Sir Howard Grubb was moved from Ireland to St Albans because they made all the periscopes for the British Navy, and these products were at hazard when sent across the Irish Sea. With the business came the key employees including Sir Howard's chief engineer Cyril Young. However, after the war the business fell on bad times and in 1925 Sir Charles Parsons bought the astronomical business and set up the new company in Newcastle, with Cyril Young as manager. The periscope business passed to Barr & Stroud at Glasgow. I joined Grubbs in 1940 as Chief Engineer under Young, a position I held until the end of the war when he retired and I took over as General Manager.

In 1925 the buildings in which Sir Charles set up the business were almost finished and had been originally intended to be a welding shop which comprised two bays; one of these now became a mechanical shop for machining, fitting, and erection of telescopes, while the other became a glass working shop and offices. The glass roof over the mechanical bay was made to roll open so that telescopes could be tested on the sky.

At the start the staff comprised six people in all. Under Young were three other ex-employees of Sir Howard: John Armstrong, who had been trained in his optical working methods and now became Optical Manager, Bob Sinclair, a skilled instrument fitter who now became Workshop Foreman, and Bill Latimer,

---

*'Planetrees', Wall, Hexham, Northumberland NE46 4EQ; formerly of Grubb Parsons, Newcastle.

a skilled optical worker who now became Glass Shop Foreman. To these were added a draughtsman from C. A. Parsons, Joe Mewett, and an office worker, Arnold Satchwell. It was not long before an order was received for a 36-in reflector for Royal Observatory Edinburgh, allowing this skeleton staff to be increased by transfer from the main works as necessary. This telescope was shown at the North East Coast Exhibition in 1929 before delivery to Edinburgh where it became the largest in use in the British Isles. This was followed by orders for a number of other major telescopes from many countries. In 1931 Sir Charles died and the Company, now employing about 25 people and privately owned by him, passed into the ownership of C. A. Parsons. Amongst his many activities, Sir Charles had been Chairman of a London optical company known as Ross (no connection with his father Lord Rosse), whose principal product was binoculars, and the Managing Director of Ross was Hasselkus. He it was who consulted for Grubbs, providing all optical designs needed for the telescope work. He was a strange, brilliant, but erratic man, very good value in conversation; one of his habits was to go down into the glass working shops and play his violin there sitting on stool — it must have been fun to see.

In 1933 Sir Charles's widow died. She had always been very interested in Grubbs activities, and being an engineer herself she had made many of the telescope parts in her works "Atlanta" which she and her daughter Rachel ran, "manned" entirely by women.

The 1930s were bad times for the engineering industry but Grubbs continued to have a steady load of telescope work. During all this stream of activity it had become necessary to test a telescope on the stars, and for the first time the machine shop roof was rolled to one side. Alas, the hot air rising from the workshop made it impossible to do any useful tests by this means; however, not to be beaten a small observatory was actually constructed behind the Works in an open field and a test mounting installed there which would carry any smallish telescope. This arrangement worked well for several years until the arrival of mercury street lighting finally put an end to its use.

The telescope series before the advent of war culminated in two orders for 74-in reflectors, the first for the David Dunlap Observatory at Toronto. This instrument was far larger than any previous telescope and presented the company with mechanical and optical problems; the crane in the mechanical workshops was limited to 5 tons, not enough to lift the 8-ton polar axis of the telescope, so the work had to be done by a laborious process involving sheerlegs and a hand-operated chain hoist. The dome was altogether too large for Grubbs to handle and was built to Grubbs design by the Cleveland Bridge Company at Darlington. Testing the 74-in mirror required the construction of a horizontal enclosed tunnel in the glass shops in which the mirror was tested on edge. However, this all went well under the control of John Armstrong; the telescope was successfully erected and set to work in Canada.

An unexpected sideline in optics was provided by an order from the Wills

Tobacco Company at Bristol for a large flat mirror 10 ft by 8 ft to be mounted as a heliostat on a building opposite the Company's northfacing board room so that sunlight would be continuously reflected onto the directors during their meetings. I wonder if it improved the profits. It certainly adds point to that well-known couplet:

> It was a summer evening old Gasper's work was done
> and W. D. & H. O. Wills were sitting in the sun.

Another surprise order on the purely mechanical side was a wind-tunnel balance for the Royal Aeronautical Establishment at Farnborough, the first of many such balances made by Grubbs. It was an elaborate type of weighing machine to measure the forces acting on a model aeroplane hanging on wires in the windstream of the 24-ft wind tunnel at the RAE in Farnborough; more of this below.

The second 74-in telescope was for the Radcliffe Observatory at Pretoria in South Africa. The negotiations with Grubbs about this telescope began towards the end of 1934, and the necessary 76-in mirror blank was ordered from Corning Glass Works in America on a delivery promise of July 1936, with an element of uncertainty about that date. How fateful that element proved to be, for an incredible succession of upsets at Corning delayed the completion of a satisfactory disc until it was finally received at Grubbs in October 1938; it had been four years in the making and no doubt cost Corning a lot more than the $10 000 paid for it. A few months later John Armstrong died suddenly and George Manville, who had joined the Company only in 1932 at the end of his apprenticeship, was put in charge of all optical working, with the immediate prospect of cutting his teeth on the Pretoria 74-in mirror now desperately needed on site. The situation gave some concern to Knox-Shaw, Director of the Observatory who consulted the Astronomer Royal, Sir Harold Spencer-Jones. As a result Young agreed that a Mr Hargreaves, an experienced amateur optical worker and astronomer, should act as a consultant during the working of the mirror. Young was quite pleased at the prospect of working with Hargreaves, feeling that he would get on better with him than he had with Armstrong who was very temperamental and uncooperative. By September 1939 the disc had been polished to a spherical surface but found to be astigmatic. At this point the Ministry of Supply directed that only war work might be carried out, so all work on the mirror was suspended. By now the telescope, complete but without optics, had been erected in the turret on site and Knox-Shaw, at his wits end, began to insist that the mirror should be sent to America and finished there. Under this threat and with the intervention of the Astronomer Royal, the Ministry of Supply reluctantly authorised a minimum effort to be devoted to further work on the mirror.

A complete list of all the telescopes made during the lifetime of Grubb Parsons is given in the Appendix.

2. Grubb Parsons: wartime

On my arrival in June 1940, work on the 74-in mirror was proceeding intermittently under the direct control of Manville, aided by visits from Hargreaves. However, this arrangement soon broke down as working with Hargreaves was a very trying process. In fact so far as Young was concerned he proved to be even worse than Armstrong, and the collaboration came to an end. In 1942 with the increased level of bombing, Young had the mirror buried under a layer of sandbags in the field behind the works to safeguard it against anything but a direct hit. He also took the precaution of lodging a plan of the burial site with the firm's lawyers, so that if all the staff were blown to bits the Observatory would know where to dig later! However, after some months the mirror was dug up again and during the rest of the war intermittent work was done on it by Manville; but it was not until 1947 that this work was finished and the mirror despatched in 1948 to complete a telescope ordered fourteen years earlier!

As assistant to Cyril Young and Chief Engineer, I had little to do with the 74-in figuring (the polishing to exact shape) until its final stages, as our work at Grubb's steadily expanded in the manufacture of small optical instruments such as aircraft identification telescopes, gunnery spotting telescopes, tank periscopes, A.A. gunsights and so on. Reorganising the production from one-off optical mirrors to quantity production was greatly helped by advice from Mr Arthur Parsons, nephew of Sir Charles, and then Chairman of Ross in Clapham, and involved many visits to London.

But apart from these optical products, an expanded range of specialised mechanical devices descended upon us. You didn't have to go looking for work then as the Ministry of Supply did a wonderful job in exploiting every potential source of engineering competence for war purposes. The Company had already made the wind-tunnel balance for the RAE, and this connection was now continued with a succession of further orders, the first of which was a six-component balance carrying the aircraft model, about 6 ft in wingspan, supported on struts in the high-speed wind tunnel. High speed was 600 mph in those days. A wind-tunnel balance is a very complicated weighing machine capable of measuring all the basic forces acting on the model aeroplane: *lift, drag* and *side-force*, as well as the torques *pitch, roll* and *yaw*. Each force has to be evaluated independently of the others and must not interfere with the readings of the other components. This makes for an elaborate mechanical assembly of levers and frames put together with consummate care, each joint being individually scraped to an exact fit. Every element in the system has to be assessed for stiffness and deflection under load so as to ensure the required accuracy of one part in 45000 called for in the specification.

A succession of further balances was made throughout the rest of the war, so that all the RAF wartime aeronautical wind-tunnel research work was done on Grubbs' products.

270

3.   Ground-based radar

In December 1941 the Air Ministry asked us to build a large rotating radar aerial which I suppose was one of the first to be made anywhere. It comprised a flat wire screen some 30 ft square and supported by a steel framework. We designed the aerial and supported it near the top on a tubular tower, and arranged a simple horizontal chain drive at the foot from a geared motor. The whole thing was drawn and built in 6 weeks, such was the urgency. I remember going down to Goring near Brighton to oversee the erection of the first unit. I went to the Mulberry Hotel at Angmering which was peaceful, but day and night the warning sirens were going for tip-and-run raiders crossing the English Channel, dumping a bomb or two on the line of hotels along the seafront at Brighton. It was bitter winter weather and the erection was much hampered by snowstorms, but we got it up and working in the end; not without a bit of friction with Merz & McLellan.

Very soon there followed a request for another aerial comprising a 30-ft parabolic reflector mounted on the front of this same framework. This had been pronounced as impossible by Merz & McLellan who acted as consultants to the Air Ministry; however, we went ahead and proved Merz & McLellan right because one of the plates supporting the extra weight did buckle a bit. However, it made no difference to the practical working of the aerial, and it was the only way the urgency of the requirement could be met. At the centre of the parabolic reflector was mounted a fearsome oscillating device carrying a 6 ft arm with a dipole at the outer end, which was not part of Grubbs' responsibility. I went down to Dunwich to see to the erection of this new aerial and was privileged to witness the first trial of the oscillating arm: it was rather spectacular because the entire arm disappeared, so rapid was the frequency, and the accompanying noise was earshattering. Somebody shouted to turn the thing off so that then we would be able to hear the bombers coming. I stayed in the Swan Hotel at Southwold overnight and for some reason was put in the Bridal Suite; next morning on entering with my early morning cup of tea the chambermaid said "Well when I came in here and saw you all alone in the bed I said to myself ho-ho there's something wrong here!" What odd memories remain in one's mind.

4.   Naval radar

I suppose the fact that we had been involved in making various radar aerials led to the Admiralty approaching Grubbs early in 1943 to take on the design of the first fully-stabilised large radar arrays intended to be fitted on the fleet carriers. These aerials, known as AQS and AQT, were large rectangular parabolic reflectors on triaxial mountings carrying their own individual gyro stabilisers.

Our design work proceeded in collaboration with the ASRE team headed by Shuttleworth based at Haslemere. Once again breakneck speed was the order of the day with many uncomfortable considerations to be taken into account, such as whether the weight would turn out too large for the intended position on the control island, and whether it was going to clear the ship's funnel. Apart from that were the mechanical problems of the drive gears free from backlash and flexibility, the arrangement of the complicated waveguide runs and so on, all to be corrosion- and water-resistant in Arctic conditions!

A complete new assembly bay had to be built at Grubbs to cope with this extra workload. Grubbs made the complete aerials as well as the gyro stabilisers but not the electrical servo driving system; these were made by Metropolitan Vickers (MV) at Trafford Park in Manchester. I remember a nasty moment when the Metadyane team from MV pointed out that their servo system might easily run out of control if a valve blew, thus driving the heavy mounting at full speed into the end of its travel. Analysis of the design showed that some 90 per cent of the momentum of the moving system resided in the motor itself, so we designed a clutch arrangement which uncoupled the motor allowing it to wind on up in speed, while simultaneously braking the actual aerial drive. An anxious moment was the first test in the workshops when the mounting was run full speed into the stops with everybody standing well back — but it worked.

Towards the end of the design a major difficulty surfaced; the electric cables on the aerials belonged to the Director of Electrical Engineering (DEE), while the mounting itself belonged to the Director of Naval Ordinance (DNO), and the clips which held the cable also belonged to DEE as did the screws which held those clips. Now DEE were standardised on BSF threads while DNO used Whitworth threads and so could not allow a BSF hole in their structure. It took a full-day meeting with twelve Admiralty representatives at Grubb Parsons to settle that, and I can't remember what we did. But *Ark Royal, Centaur, Albion,* and all the fleet carriers were fitted with these aerials, which continued in production for some years after the end of the war.

5.   Astronomy comes to life again

During all this time the veto on any astronomical work had held good, but by the end of 1944 first contacts with the astronomers had been made. I was now in touch for the first time with the Astronomer Royal, and also H. H. Plaskett, Director of the Oxford University Observatory, who was planning to build a Solar Telescope. And so during 1945 my introduction began to what was really going to be my life's major interest in the years after the war.

By 1944 the danger from bombing had become so reduced that the 74-in mirror was dug up and Manville started the final stages of figuring, the ultimate polishing stage when the surface has to be smoothed out to a parabolic shape with an accuracy of about one millionth of an inch. All the testing is done by use of optical systems which project beams of light at the mirror and then measure where the reflected light has returned. In those days the mirror was supported on edge in a test tunnel where the air could be kept still and at a uniform temperature. Figuring is a strange process, because after a polishing run on the mirror surface lasting perhaps ten minutes, the mirror must cool and the air-column stabilise; it may take a week or more to be quite sure what alteration you have made to the shape.

As it became clear that the war was ending, I was able to discuss the new Photographic Zenith Tube (PZT) telescope for the Royal Observatory; in addition the Astronomer Royal had agreed plans with the Admiralty to move the Observatory from Greenwich to a clearer air, for which purpose the Castle of Herstmonceux in Sussex had been selected. Nearly all the instruments from Greenwich were to be removed to the new site, and new domes were to be built for them; Greenwich would become a museum. Strange that as I write the same fate has overtaken Herstmonceux as the RGO has been moved yet again, this time to Cambridge. The Astronomer Royal was starting to receive enquiries from Observatories abroad about the prospect of telescope manufacture starting up again in England; so now for the first time I met Sir Harold Spencer-Jones with whom I was to have so many contacts over the years; he was so tall and seemed so remote and pompous that I felt rather daunted. But this was appearance only, as in fact he was immediately most kind and helpful, proposing me for membership of the International Astronomical Union (IAU).

In a meeting he had a disconcerting habit of sitting looking out of the window at the swans on the moat all the time you were expounding a complicated matter, so that after a bit you wondered if you were making any contact at all. But again there was a surprise because you found he had in fact a complete grasp of what you had said. He was a diplomat and thought like lightning; I remember early in our acquaintance I asked him what a comet was made of. Like a flash he replied "An aggregation of cometaceous material". On another occasion he was giving a public lecture to the Newcastle Astronomical Society on the Origin of the Universe, and at the end during questions an old lady got up at the back with a small quavering voice to ask "Don't you think these signals from outer space are really the voice of God?". Instantly he replied "I don't see how the voice could travel through a vacuum".

At about this time also the Admiralty Research Laboratory asked us to do some work on infra-red detection of enemy ships and vehicles, and following this contact they also asked us to make an infra-red spectrometer to the design of Dr A. Elliott, who headed their team. This introduction to the infra-red instrumentation field was subsequently to prove very important to us.

6.   Peacetime – picking up the strings: 1945

With the end of the war Cyril Young retired from Grubbs and I took over the management of the company from him with George Manville as my assistant. We were in full flood with the radar contract for the Admiralty; but the end of that was in sight, and our military work was clearly going to end soon. Here were a couple of hundred people all up to their eyebrows in work but quite soon they would have nothing to do unless we could get busy with some astronomical work, enquiries for which were starting to come in. Even so it was clear that this was not going to fill our present capacity, and it was my job to find something to keep us busy. We had a meeting with the Admiralty to ask about future prospects and after a visit of inspection to Newcastle they informed us that we were just a bunch of mechanical engineers, and so far as they were concerned it must be thank you and goodbye. Now our brief contact with infra-red apparatus led to a visit from Professor H. W. Thompson (later Sir Harold) of St John's College, Oxford. He encouraged us to enter this field against competition mostly from Perkin-Elmer in America, and he suggested that we engaged Dr A. E. Martin from the Government Chemist's Laboratory to lead a technical team. So it came about that Dr Martin joined us and remained with the firm until he retired in 1969.

7.   Infra-red instruments: 1946

As a result of Professor Thompson's visit a completely new field of infra-red instrumentation came to be made by Grubbs, starting with spectrometers and Luft-type gas analysers. The first spectrometers were double-beam multiprism types and by 1950 these were in quantity production. But by then cheap coarse gratings had become available by the development of the Merton process at the NPL. The first commercially-available grating spectrometer made by Grubbs, the GS2, was a world leader for several years, and was followed by a series of further models. The major contribution to these was by J. Shields who was responsible for the detailed designs. Later, in collaboration with J. Gould of Reading University, he extended his activities to include a milk analyser. Interferometric spectrometers followed, stimulated by the work of A. Gebbie at the NPL. The scientific instrument market was very competitive with the American firm Perkin-Elmer our fiercest rivals, not only in the infra-red but also in telescope manufacture. It was a year or two before we discovered that when we both exhibited at the annual SIMA Instrument Exhibition they used to trawl our waste paper basket in the evening to monitor any scraps of useful information! Alongside the spectrometers a series of gas analysers was produced by a team under the leadership of John Smart. All this activity kept about half of Grubbs capacity going, and fitted in quite well with the astronomical work.

Meanwhile what of telescopes? – With Cyril Young retired there was almost nobody left who had ever seen a telescope project through. The management of our owners C. A. Parsons had changed with the arrival of Sir Claude Gibb as their new Managing Director, and fortunately he took a very sympathetic line towards us; in fact he authorised a considerable extension of our premises to accommodate the infra-red work.

Manville was now at last able to turn his full attention to the final stages of the Pretoria 74-in mirror which had a zonal error near its edge, the significance of which was baffling everyone. A skilled amateur optical worker, Dr Steavenson, was acting for the Observatory as an accepting authority, and was prepared to accept the mirror; but the Astronomer Royal thought that a second opinion was desirable and asked Professor C. R. Burch at Bristol University to give it his attention. The unforeseen importance of this event was that one of Burch's students, W. J. Bates, invented the waveshearing interferometer and used it for the first time on the Pretoria mirror with immediate success. The device with subsequent modifications was used at Grubbs ever after as the major test weapon. In various capacities Burch over the years became a firm friend of the family; he was a tiny wizened figure and immensely wise — when we played the dictionary game he always won easily with his knowledge of Greek and Latin to help him. Whereas we reckoned that my wife and I knew the general meaning of about 17 000 words he regularly clocked up well over 20 000.

My first astronomical gathering was the Tycho Brahe centenary celebrations in Copenhagen that year. It was a very good party; I had been forewarned about the need for a tail suit so that I had mine with me. The King appeared to enjoy himself so much that he simply would not leave, and soon everybody was in despair as nobody could go until he did. In the end etiquette was sacrificed and I as well as a few more crept out.

8.   Tendering problems

We were now in the throes of quoting for a number of telescopes ranging from two 7-in Meridian Circles to three 74-in reflectors, and several others. We had the drawings of some of these instruments as made before the war, but now many new ideas were to be incorporated. We also had the cost figures from pre-war; for instance we knew that Cooke Troughton and Simms had quoted the Royal Greenwich Observatory £4950 for a 7-in Meridian Circle shortly before the war and it had cost them £7487 to make.

We didn't want to repeat that sort of thing. Our price for the Meridian Circle estimated in 1947 now came out at £19 050 and by the time it was delivered inflation increases specified in the contract had increased that to £24 427 on which we made a small profit. So the difference between pre- and post-war costs was huge.

So far as big telescopes were concerned much more money was at stake and we were busy estimating for 74-in reflectors for France, Mt Stromlo in Australia, and for Cairo. Our first approximation submitted to Cairo University for their 74-in unfortunately contained a serious underestimate of this factor, and a few months after its submission I had to go out to Egypt with the bad news that the real price must be a lot higher than originally given as our approximate price. Of course this placed the Professor of Astronomy, Madwar, in a difficult position but no way could I revert to the original estimate.

The impact of Cairo was naturally terrific. I stayed with our agent William Long, and all night the cats in the courtyard below my bedroom window vied with the motor horns making sleep unlikely. Day and night the motors in the narrow streets blew their horns in a wild cacophony — I suppose it is still the same. Our meetings at the University were a bit tense, and in a last throw to move me Madwar suddenly said "But Sisson this is a Royal Observatory you understand. Tomorrow I will take you to the King and you will have to agree with him. Oh Sisson you could be a Pasha!". However, I reckoned it would cost our Company too much for me to acquire this honour and refused — perhaps I would get one at the end, but come that time things were very different in Egypt!

The whole question of tendering for telescopes was always difficult, even after we got a bit of experience. You seldom knew quite what the engineering was going to involve, and there was competition from several other makers so it was by no means an open field. Some telescopes were similar but others were something completely new, and of course you had to get it right first time. A potent factor in the estimate was my assessment of the customer, and just how fussy he was going to be over details when we came to flesh out the original specification. Then there were Contract Conditions. Our first big tender was to the French Government for the St Michel 74-in reflector. Of course I went to consult our bosses in C. A. Parsons where the Sales Director Harold Martin was a real hard nut; he made contracts for power stations worth many times more than Grubbs annual turnover. "We sell on BEAMA Conditions of Sale AE. They have stood the test of time and you must sell on them; do not alter one word." In vain did I point out that turbine sales might be a little different from telescopes, and that items referring to guaranteed steam consumption might appear mysterious to an observatory.

All to no avail, so I tendered to France as ordered.

Very soon I found myself in Paris confronting a very amused senior civil servant, and desperately trying to make our tender sound even meaningful. However, absolutely no problem as with impeccable logic he extracted the essential elements and we agreed a rewording which satisfied both of us; this then became our tender and remained so for every other job we took on. I never said a word to Harold Martin and he still doesn't know we did it.

9. The International Astronomical Union

My membership of the IAU was inaugurated by a visit to the first post-war Congress in July 1948 at Zurich — I felt rather nervous, having no idea of what was in store; but at least I knew that I would meet many friends in the astronomical world with whom I had already had discussions about various projects. On the aeroplane the very lovely air hostess came and sat next to me and poured out her lament that while in Zurich she would have two lonely days to kill with nothing to do and what a bore it would be. I said to myself, now Sisson, resist; I did and she went and sat next to somebody else. I lodged as a guest with a lawyer's family in the suburbs.

I suppose about 400 people assembled for that meeting, and at the final dinner some unexpected excitement occurred. The Cold War was in full swing and the Russian delegation had not replied to their invitation to the Congress; however, they turned up in force a few days late and attended the rest of the sessions. During the farewell dinner Donald Sadler, who was sitting next to Bill (later Sir William) McCrae, noticed that amongst the forest of national flags decorating the end of the dining room no hammer and sickle was to be seen; he unwisely sent a message to the President which alerted the Russian Chief Minder who rose to his feet and commanded in a loud voice that all should leave forthwith. The unfortunate USSR astronomers scattered around the tables got up and filed out. Shortly workmen appeared carrying a pole with the red flag upon it, set it amongst the rest with much hammering, and the Russians then returned to finish their dinner.

I attended every Congress after that, as a representative on two Commissions; it was a wonderful source of new friendships worldwide.

10.   Conclusion

The end of my direct executive responsibilities in Grubbs came when in 1969 I was asked to take control of a newly created Technology Division in Parsons which included Grubbs as a very small element. A few years later in 1974 I severed all connections with Parsons, including sadly Grubbs. The team which I had left in place at Grubbs continued to make telescopes for another ten years, including some of the largest ever made by the Company. The optics team headed by David Brown with David Sinden doing the actual glass figuring was superb, and the mechanical side was in the care of Munro and Andrew Laurie.

However, the Isaac Newton was the last large telescope actually designed by the Company and the subsequent ones were made to given designs. I do not doubt that the competitive element inherent in this procedure sharpened up their tendering for I know that serious losses were incurred on those later telescopes. Misguided efforts to reduce costs led to the discharge of Sinden who set up his own optical business; this left David Brown on his own, and I am glad to say he was completely successful with the La Palma optics. With the infra-red instrument side also falling into trouble, from the purely commercial point of view, the end could not be avoided because telescope making on its own was not a viable business and the thriving instrument activity had to complement the astronomical work.

Telescope building is a succession of adventures and I have vivid memories of many on the long list in the Appendix, except the last few. It would be impossible to relate them all here, but they are being recorded and the comments contained above have been extracted from that history and so may appear a little sporadic. The Appendix gives a brief outline of Grubbs history; some of the telescope completion dates are approximate. The scientific side of our products was always changing and it was a battle of wits to keep ahead of the competition and choose the right things to do next; I had complete managerial freedom to do what I liked except make losses. Looking back on it all I cannot imagine I could have had a more enjoyable lifetime's work.

For additional information ...

References

[1] Ball Sir Robert, *Extracts from the Diary of Sir Robert Ball. The Observatory, Vol. 63 (1940), pp. 795, 797.*

[2] FitzGerald W., *Sir Howard Grubb. Strand Magazine, (Oct. 1896), p. 369.*

[3] Gascoigne S.C.B., Proust K.M. and Robins M.O., *The Creation of the Anglo-Australian Observatory. (1990), Cambridge University Press.*

[4] Glass I.S., *The Story of the Radcliffe Telescope. Q. J. Roy. astr. Soc., Vol. 30 (1989), p. 33.*

[5] Graham-Smith F. and Dudley J., *The Isaac Newton Telescope. J. Hist. Astron., Vol. 13 (1982), p. 1.*

[6] Manville G.E., *Heaton Works Journal, Vol. 11 (1967), No. 64.*

[7] Sisson G.M., *Sir Howard Grubb, Parsons and Co. Proc. R. Soc. A., Vol. 230 (1955), p. 147.*

[8] Sisson G.M., *Parsons Memorial Lecture. NE Coast Inst., (Nov. 1968)*

[9] Sisson G.M., *Obituary, David Brown. Q. J. Roy. Soc., Vol. 30 (1989), p. 279.*

[10] Sisson G.M., *Design of Large Telescopes. Vistas in Astron., Vol. 111 (1960), p. 92.*

APPENDIX: Summary of Grubb Parsons Telescope Manufacture

1925   Foundation of Grubb Parsons by Sir Charles Parsons at
Walkergate, as his own private company. Six employees,
Manager Cyril Young.

1925--1931

Sir Charles negotiates with Pulkova Observatory for a 41-in
refractor but it came to nothing.

36-in reflector	Edinburgh
40-in reflector	Stockholm
24-in/20-in refractor	Stockholm

1931   Sir Charles dies. Grubb Parsons now with 25 employees,
ownership passes to C. A. Parsons. Sir Howard Grubb also dies.

1931--1939

36-in Reflector	Royal Greenwich Observatory
15-in/10-in Twin Refractor	Edinburgh
16-in Refractor	Leiden
16-in Solar Telescope	Oxford
18-in Reflector	Dundee
13-in/10-in Twin Refractor	Poland
74-in reflector	Toronto
10-ft x 8-ft heliostat mirror	to provide sunshine into the Board Room of the Wills Tobacco Factory at Bristol
74-in Reflector	Pretoria. The mirror of this telescope was caught by the war and was not finally delivered until 1948.

1940--1945 G. M. Sisson Assistant Manager. Variety of war work:
 radar aerials, wind-tunnel balances.

1945--1969

G. M. Sisson General Manager and Chief Engineer.
G. E. Manville Optics Manager.
About 200 employees.   Infra-red instruments started up.

Date	Aperture	Telescope type	Observatory
48	12-in	Solar Telescope	Arosa
50	20	Solar Reflector	Oxford
52	74	Reflector	Mt Stromlo
53	7	Transit	San Fernando
54	17/24	Schmidt	Cambridge
54	10	Photo Zenith Tube	Neuchatel
54	7	Transit	Copenhagen
55	36	Reflector	Cambridge
55	10	Photo Zenith Tube	RGO
56	10	Photo Zenith Tube	Mt Stromlo
56	75	Reflector (no optics)	St Michel
58	24	Coelostat	Kodaikanal
60	74	Reflector	Tokyo

1961   D. S. Brown appointed Optical Manager,
        D. Sinden Optical Workshop Manager

61	48	Reflector	Victoria
62	16	Reflector	Edinburgh
63	40	Reflector	Capetown
63	74	Reflector	Helwan
66	30	Reflector	Jungfraujoch
67	20	Reflector	Edinburgh
67	16/24	Schmidt	Castel Gandolfo
67	98	Reflector	RGO
68	20	Reflector	Glasgow

1969   G. M. Sisson Director C. A. Parsons, Technology Division,
Chairman of Grubb Parsons (non-executive). Grubb Parsons
employees about 250. J. B. Price new Managing Director.

1969--1985

Date	Aperture	Telescope type	Observatory
73	48	Schmidt (UKST)	Siding Springs
74	160	Optics and tube (AAT)	Siding Springs

1974   G. M. Sisson retires from C. A. Parsons and Grubb Parsons.
1975   D. S. Brown appointed Technical Director Grubb Parsons.

75	48	Optics and tube	Hamburg
75	48	Reflector	Athens
75	30	Optics and tube	S Africa
76	150	IR thin mirror (UKIRT)	Hawaii
76	60	Optics and tube	ESO

1976   D. Sinden leaves Grubb Parsons to start own company.

77	31	Schmidt mtg (no optics)	Max Planck
79	39	Reflector (Jacobus Kapteyn)	RGO La Palma
79	30	Reflector	Portugal
80	100	new mirror, Isaac Newton	RGO La Palma
85	170	Reflector (William Herschel)	RGO La Palma

85   Grubb Parsons closed down by NEI.

23

Charles Wynne – an Astronomer's View

Gerard Gilmore *

Abstract

Astronomical instrumentation requires both innovative technical development in university departments, and design, construction and maintenance of proven systems for common use on national facilities. Charles Wynne has contributed substantially to both types of instrumentation throughout his career, and continues to contribute today. Some examples of his work in an historical context are given, together with some historical lessons for present funding and research strategies.

1. Preamble

Before beginning an astronomer's view of this meeting, and the work of the person we are honouring here, let me first say what is wrong with this meeting, and what Charles Wynne has done wrongly since coming to Cambridge.

You will be aware that funding for the Royal Greenwich Observatory is based on a five-year cycle of Forward Looks, with it must be said startlingly little consistency on a five-month cycle within that period. None the less, meetings such as this 80th birthday celebration must be planned in the Forward Look, up to five years ahead. So already someone has pencilled in an 85th birthday, ... Be warned! One of the less well understood aspects of Snell's law is the longevity it awards to those who really understand it. Charles Wynne, sprightly youngster that he is, can confidently expect to become a permanent feature of the 5-year Forward Look unless these meetings become 10-year, or even 20-year events!

There is a similar phenomenon in stellar dynamics, where the 'Oort limit' is now being redefined as the number of five-year celebrations which can be achieved after retirement. This 'Oort limit' – a large number – is potentially under threat from the 'Wynne number', if Charles's present vigour and productivity are anything to go by.

Now to Charles's faults. There is in my department a prominent scientist who believes all Cambridge professors become dotty in their 50s. I shan't tell you his name, but can say that he has recently resigned as Plumian Professor when reaching 50 – a rare example of a cosmologist apparently believing a theory enough to act

*Institute of Astronomy, Madingley Road, Cambridge CB3 0HA, UK.

on it. Charles Wynne is manifestly not dotty, and thereby violates local custom. Mistake number one.

There is another aspect of Cambridge astronomy which is at least as well established as the Rees rule above – one should work on subjects which are neither understood, nor explicable by classical physics. Charles has mastered physical optics. Mistake number two. In the rest of this paper I will give some examples to illustrate how fortunate the astronomical community is to have these two mistakes, and to have Charles Wynne.

Before starting, however, it is useful to remind oneself of the state of astronomical optics before those at this meeting took the subject in hand. Fig. 1 shows a Heath Robinson illustration from *The Sketch* of 17 November 1909, showing astronomers 'observing' Halley's comet. Noting that the date of publication is nicely nine months before Charles's birth, one wonders if Mr and Mrs Wynne decided to do their bit for modern British astronomy. Whether deliberate or not, they succeeded.

2.   Astronomical instrumentation – who built it?

The most important feature of any astronomical instrument is having an astronomer who wants to use it. Charles Wynne often says: "I don't want just to design optics. I want to design optics which people want to build." However, I think this is only partly true.

There are *two* ways in which astronomical technology progresses, and it is essential for the viability of the subject that both are in a healthy state. These two complementary aspects are the design and development of special-purpose or prototype instruments, and the provision of fully engineered common-user systems.

Special-purpose instruments have been described in several excellent presentations at this meeting, ranging from the UV work of Conal McKeith, through the high efficiency, high resolution studies described by John Meaburn, based on optics by Charles Wynne, to the outstanding performance of the narrow-band Fabry–Perot imaging system TAURUS of Richard Bingham. It is clearly not chance that *all* these prototype instruments exploited state-of-the-art optical (and electronic) designs to produce state-of-the-art science. It is also clearly not chance that everyone was initiated and developed at least in large part in a University department by people with an astronomical motivation.

This illustrates the first essential aspect of progress in experimental science. Somewhere in a University department a scientist has an idea, and wants to test it. Where the technical design and construction skills also exist, substantial scientific and technical progress can be made. It is interesting that it is cross-fertilisation between several (small) groups in more than one department – typically involving Charles Wynne in optics, together with electronic and computing contributions – which has produced the most successful innovations.

The limiting examples of this scale of innovative technical research are the three projects discussed here, COAST by Peter Warner, COME-ON (Fritz Merkle),

Science Jottings — By "Dr." W. Heath Robinson (D— L—).

I.— SEARCHING FOR HALLEY'S COMET, AT GREENWICH OBSERVATORY.

Figure 1. From *The Sketch*, 17 November 1909: astronomers and Halley's Comet.

and SUSI (Hanbury Brown). Each of these is a remarkable example of scientific and technical innovation, each is motivated by scientific research goals, each is a collaboration between groups, each is a substantial advance in high spatial resolution imaging, and each is a University-based project. Each is also at the limit of scale which can be managed within a single research group.

There are two lessons to be learnt from this. The more obvious is the need to ensure a healthy programme of instrumental development in research groups – a simple lesson which I hope is not lost on those determining funding policies. The other lesson is derived from the limitations of the present systems. Projects such as COME-ON, the adaptive optics project, were feasible only because of the involvement of several groups, and commercial aerospace interests.

The scale of modern astronomical instrumentation, particularly that required to exploit fully the new generation of 8m telescopes, is too great for single groups. It is time for astronomers to learn from particle physicists, and establish consortia of universities to develop state-of-the-art instrumentation. Adaptive optics seems to me a perfect example when such supra-institutional, and probably supra-national, efforts are going to be needed. High spatial resolution is clearly the next field of rapid scientific advance in astronomy, and equally clearly is not going to be mastered at the technological level by a transient or underfunded group. It will be interesting to see if the community can rise to meet this challenge.

Following successful development of a prototype instrument, there is a need to allow the maximum number of ideas and discoveries to be tested or made with it. This of course requires access by astronomers outside the group of builders, and so presumes *common-user* facilities. There is a significant difference in the skills and resources required to develop and maintain an engineered system from those required to develop prototypes. Hence, for efficiency and cost-effectiveness, one has a two-tier system of facility development, with national facilities complementing the research and development. Neither tier is sustainable without the other. Some subset of new instruments are of wide general use, and those need to be chosen, engineered, built, commissioned, maintained, and supported for use by scientists – the majority – with good ideas but no detailed ability to manage an instrument themselves. This role is carried out most efficiently by the national observatories, currently managed by RGO and ROE, both of which have benefited from Charles Wynne's expertise over the years.

3.   Astronomical instrumentation – who designed it?

Astronomers are too ignorant to predict new discoveries. Rather than progress in understanding by pure thought, *á la* Descartes, astronomy still progresses by Baconian empiricism. For that we need telescopes, and for those we need optical designs.

It is interesting to look a little at the history of the large telescopes built this century to see why observational astronomy has progressed so well, and to see who deserves credit for this progress. The most famous telescope is certainly the Palomar 200-in reflector. Bowen showed that the design for this should be matched to the best available detector, the IIaO plate, so that f/8 is required. This in turn requires a Ritchey–Chrétien design at the Cassegrain focus. No prime focus was then available. Mayall realised the desirability of access to the prime focus, and initiated a search for a generalised (Ross) corrector to allow this access. This led eventually to the Wynne corrector, from which *all* large-telescope prime-focus correctors are derived.

Advances in photographic technology, and later CCD's, made obsolete the IIaO design constraints, so it is interesting to see why the 200-in did so well. Fundamentally of course, it failed! One of the primary scientific goals for the 200-in was detection of the RR Lyrae variables in M31, to calibrate Hubble's and Baade's distance scale. It has yet to achieve this, though the experiment was eventually achieved in the 1980s by the Canada-France-Hawaii Telescope. Equipped with a Wynne prime-focus corrector.

So, given the irrelevance of its technical design constraint, and the failure of its notional scientific goal, why is the 200-in a scientific success? In large part this is due to the Wynne optics. These provide a sufficiently good general purpose facility that clever people can do experiments not dreamed of at the time of its design. So, another good lesson for today – don't 'save' money by removing features from telescopes, and don't attempt to make or follow defined scientific plans. Rather provide technical resources for clever people, stand back, and watch progress.

There is another interesting piece of history in this story, and that is why it was that Charles Wynne was the person who solved the prime focus corrector problem. Part of the answer is of course easy – he is very clever. The other part is that his group at Imperial College had the most modern technology, and used the most innovative methods. To understand why that should be, a brief look at the history of optical design in the 1940s and 1950s is needed. A somewhat longer history can be found in the December 1963 edition of *Applied Optics*, while a forthcoming book by R. N. Wilson – a former colleague of Wynne and himself a distinguished designer of optical systems – contains a more comprehensive history.

Modern optical design in astronomy almost started in 1941 with a proposal by R. Minkowski to the National Defence Research Council that automated least squares calculators should be used to investigate possible optical designs. He was ignored. By 1944 Baker at Harvard was using the Mark I calculator for ray tracing. With this device a single ray could be traced through one surface in a mere 120 seconds!

By the early 1950s such calculations were sufficiently common that the Air Force became interested. They funded a detailed study by Baker from 1951–1955 through the Perkin-Elmer Corporation – a company now notorious for its modern astronomical optics. This report was classified until 1959, while Baker's grant application for a faster machine to contine his work was rejected. This combination of classification rather than open dissemination and lack of support for forefront computing gave the lead in optical design to the UK. One trusts this historical lesson will continue to be heeded. Certainly Charles Wynne learned it, as a consistent policy to make his designs public has been a commendable part of his career, and a major contributor to his reputation.

In the UK, Black in Manchester had published optical designs carried out in the early 1950s with a Ferranti calculator. I assume this was the origin of the interest shown in the subject by P. M. S. Blackett. In any case, when Blackett moved from Manchester to Imperial College, London, he brought to Imperial (from Wray Optical Works) the young(ish) Charles Wynne.

By that time Wynne was already known for his innovative design work, his open publication policy, his mathematical rigour – his damped least squares algorithm was christened SLAMs, showing that acronyms have not improved since the 1950s even if computers have.

In the 1963 *Applied Optics* issue, Donald P. Feder wrote of Charles Wynne: "Before becoming interested in automatic computers, Wynne had already had considerable experience in optical design, and he has a very clear physical picture of the processes involved which, coupled to a facility for expressing himself clearly, has made his work very accessible and easy to read." It is this combination of rigour, openness, and an appreciation of what surfaces can actually be built, which has marked out the designs we all use today.

Feder continued: "[Wynne] hopes shortly to acquire an extremely fast machine, and we may expect that he will turn out noteworthy designs at an accelerated pace". He did and he did.

That same issue of *Applied Optics* contains a paper by Wynne and Wormell with a generous acknowledgement to a young designer named Ray Wilson. This nicely illustrates the extent to which the generation of optical designers trained by Charles Wynne's group at Imperial College has continued, and is continuing, to dominate optical design today. A dominance to which the teacher and the pupils continue to contribute.

The UK and the astronomical community is fortunate that short-sighted (lack of) publication policy by the US Air Force in the 1950s, together with relatively good funding of innovative research at Imperial College, allowed the IC optics group to flourish.

Before looking forward, it is perhaps relevant to note that the Wynne corrector in all the current large telescopes is not the only contribution by Charles Wynne and his group to modern astronomical instrumentation. In addition to many spectrographs and cameras, some few of which are mentioned in these Proceedings (and 41 patents!) there are also the wide-field Schmidt designs. Large telescopes of course have relatively small fields of view. To use them efficiently one needs to know in advance what objects are likely to be of interest. Usually this information comes from a sky survey in some bandpass – radio, infrared, X-ray, or whatever. The optical sky surveys are provided by three excellent Schmidt telescopes – the

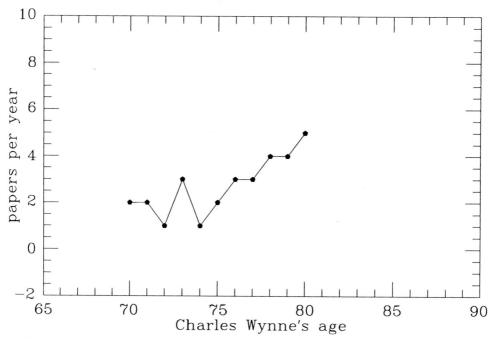

Figure 2. Charles Wynne: annual publications since "retirement".

Oschin/Palomar Schmidt, the ESO Schmidt, and the Anglo-Australian Schmidt. All three use a Wynne-designed corrector.

4. Charles Wynne – an astronomer's view

I first met Charles and Jean Wynne when they came to Cambridge. Though both in their mid-70s, their intentions were clear from the start by the vigour with which twitch and ground elder disappeared before their advances. Academically, rather similar behaviour has been apparent, as the recent rate of publications by Charles shows. Fig. 2. shows the annual rate of publication of papers since retirement. Simple extrapolation to age 90 is awe-inspiring!

So, after all, what is an astronomer's view of Charles Wynne. We owe Charles the prime-focus field on our large telescopes. We owe him the survey telescopes. And we owe him many beautiful, efficient, and effective spectrographs and cameras. The whole of modern observational astronomy has been profoundly improved by Charles Wynne's optical designs. What else can one say, but Thank You.